Leadership Matters

This book presents in-depth case studies from an industrial engineering perspective of the rise and fall of selected industries around the world. Discussing what works and what doesn't work from a systems approach can help the industry avert problems of shutdowns and takeovers.

Leadership Matters: An Industrial Engineering Framework for Developing and Sustaining Industry discusses how industries face the birth and death of new manufacturing enterprises. Many disastrous outcomes can be prevented if proper industrial engineering tools and techniques are applied throughout daily operations. This book explains how product development, advancement, and international promotion are best accomplished through the practice of industrial engineering techniques. It illustrates how the discipline of industrial engineering has a proven track record of aiding the survival of industry especially now that organizations are more vulnerable to adverse developments in the global supply chain. Real-world examples of managing products, services, and results are offered along with a range of management tools. A template for operational excellence is also provided.

This practical professional book should be on the reading shelf of all industry practitioners, industrial and manufacturing engineers, business management professionals, and consultants as well as those teaching industrial, mechanical, and business leadership courses.

T0171833

Systems Innovation Book Series

Series Editor: Adedeji B. Badiru

Systems Innovation refers to all aspects of developing and deploying new technology, methodology, techniques, and best practices in advancing industrial production and economic development. This entails such topics as product design and development, entrepreneurship, global trade, environmental consciousness, operations and logistics, introduction and management of technology, collaborative system design, and product commercialization. Industrial innovation suggests breaking away from the traditional approaches to industrial production. It encourages the marriage of systems science, management principles, and technology implementation. Particular focus will be the impact of modern technology on industrial development and industrialization approaches, particularly for developing economics. The series will also cover how emerging technologies and entrepreneurship are essential for economic development and society advancement.

Global Manufacturing Technology Transfer: Africa-USA Strategies, Adaptations, and Management
Adedeji B. Badiru

Data Analytics: Handbook of Formulas and Techniques
Adedeji B. Badiru

Conveyors: Application, Selection, and Integration
Patrick M McGuire

Innovation Fundamentals: Quantitative and Qualitative Techniques
Adedeji B. Badiru and Gary Lamont

Global Supply Chain: Using Systems Engineering Strategies to Respond to Disruptions
Adedeji B. Badiru

Systems Engineering Using the DEJI Systems Model®: Evaluation, Justification, Integration with Case Studies and Applications
Adedeji B. Badiru

Handbook of Scholarly Publications from the Air Force Institute of Technology (AFIT), Volume 1, 2000–2020
Edited by Adedeji B. Badiru, Frank Ciarallo, and Eric Mbonimpa

Project Management for Scholarly Researchers: Systems, Innovation, and Technologies
Adedeji B. Badiru

Industrial Engineering in Systems Design: Guidelines, Practical Examples, Tools, and Techniques
Brian Peacock and Adedeji B. Badiru

Leadership Matters: An Industrial Engineering Framework for Developing and Sustaining Industry
Adedeji B. Badiru and Melinda L. Tourangeau

Leadership Matters
An Industrial Engineering Framework for Developing and Sustaining Industry

Adedeji B. Badiru
Melinda L. Tourangeau

CRC Press
Taylor & Francis Group
Boca Raton London New York

CRC Press is an imprint of the
Taylor & Francis Group, an **informa** business

Designed cover image: Shutterstock

First edition published 2023
by CRC Press
6000 Broken Sound Parkway NW, Suite 300, Boca Raton, FL 33487-2742

and by CRC Press
4 Park Square, Milton Park, Abingdon, Oxon, OX14 4RN

CRC Press is an imprint of Taylor & Francis Group, LLC

Library of Congress Cataloging-in-Publication Data

Names: Badiru, Adedeji Bodunde, 1952- author. | Tourangeau, Melinda L., author.
Title: Leadership matters : industrial engineering framework for developing
and sustaining industry / Adedeji B. Badiru, Melinda L. Tourangeau.
Description: First edition. | Boca Raton, FL : CRC Press, 2023. | Series:
Systems innovation book series | Includes bibliographical references and index.
Identifiers: LCCN 2022059507 | ISBN 9781032317809 (pbk) | ISBN
9781032317861 (hbk) | ISBN 9781003311348 (ebk)
Subjects: LCSH: Industrial engineering. | Transformational leadership. |
Corporate governance. | Corporations--Growth.
Classification: LCC T56 .B285 2023 | DDC 620.0068--dc23/eng/20230113
LC record available at https://lccn.loc.gov/2022059507

ISBN: 978-1-032-31786-1 (hbk)
ISBN: 978-1-032-31780-9 (pbk)
ISBN: 978-1-003-31134-8 (ebk)

DOI: 10.1201/9781003311348

Typeset in Times
by KnowledgeWorks Global Ltd.

Dedication

Dedicated to leadership and partnership and the power of believing.

Contents

Preface

Industries come and go. It is exciting when a new industry develops and thrives. It is disheartening when a thriving industry collapses and is taken over by a competitor. This book presents an in-depth guide, based on the tools and techniques of industrial engineering, to grow and sustain industry for the sake of national industrial development goals. All nations depend on industry to generate and distribute products and services to meet the needs of the population. Far too often, we see industries come up in a flash, only to diminish in a crash. Effective leadership is the bastion of the growth and sustainment of industry. Thus, the premise of this book centers on how leadership matters and what matters in leadership in the pursuit of industry advancement and survival. In this context, "industry" is the collective pursuit of producing end products from raw materials. This includes all types of manufacturing activities. The typical outputs of an organization fall in three categories:

1. A physical product (e.g., consumer products, food items)
2. A service (e.g., a social service)
3. A desired result (e.g., voter registration)

The genesis of this book is that leadership matters and is a matter of concern in organizations. It was as long ago as 2002 that two consulting partners, Mr. Tony Mayfield and Mr. Jason Chandler, accompanied the lead author on an industrial training program for Honeywell Flour Mills (HFM) in Lagos, Nigeria. Visibly impressed with what they observed at HFM, they floated the idea of writing a formal corporate case study on Mr. Tunde Odunayo, the prevailing managing director of HFM at that time. They toyed with the idea for many years but never got around to writing the case study. It was the sudden takeover collapse of HFM in 2021 that served as the last straw for Adedeji B. Badiru to partner with Melinda L. Tourangeau to get the formal case study written at this time, with an expansion to include a general industrial engineering foundation for leadership as well as other case examples from business, industry, academia, or the military.

Case studies as well as foundational principles are used to convey the message of this book. Of particular focus is the rise and fall of HFM, although other cases will be used as well. Although this international case study is situated in Nigeria, the lessons learned are pertinent and useful for all industries, particularly those in developing countries, including those in the evolving markets of the United Kingdom, the European Union, and Asia. The COVID-19 pandemic era makes this case study approach even more relevant and needed in response to the global industrial supply chain disruption and the ensuing scramble for industry survival.

Adedeji B. Badiru
Melinda L. Tourangeau

Acknowledgments

We acknowledge the support and contributions of our friends, colleagues, and family in the development of this manuscript. Several people, too numerous to name individually, helped along the way. Of particular recognition is the leadership and support provided by Ms. Cindy Carelli, our Executive Editor at CRC Press/Taylor & Francis, and her team. She and her team, including Ms. Christina Graben, demonstrated exceptional responsiveness in attending to all our administrative and editorial needs.

Preamble

The rise of Honeywell Flour Mills (HFM) under the leadership of the founding managing director, Mr. Tunde Odunayo, epitomizes the spirit of industrial engineering through Odunayo's resourceful application of industrial engineering principles while carrying out his proactive leadership style. His actions are a demonstration of how *leadership matters*, as conveyed by the title of this book.

Having followed the industrial growth and corporate success of HFM since 2002, through direct industrial consulting for the company it was with a great shock, disbelief, and disappointment that the lead author read about the corporate demise of the nationally beloved company in November 2021. The template of success, which Mr. Odunayo created, had been expected to be an industrial standard that others could emulate. Instead, what we saw was how the corporate entity had been decimated by subsequent poor leadership, wrong choices, and administrative ineptitude within a few years after the retirement of Odunayo in 2015. Even in a pseudo-perfect organization, a well-laid-out leadership succession plan would have saved the company and sustained it on its well-known track of advancement. It is for this reason this book has been written to capture and preserve the best practices of this pioneering industrial standard, in the hopes that leader-readers will see its worth and embrace its spirit for their own organizations.

The diverse contributions of Tunde Odunayo, although in a regionalized market, typifies the accomplishments of major industry executives around the world. When looking for case studies and industrial leadership examples, writers, authors, and scholars should open the aperture of case-study reviews beyond typical Western companies. There are good gems and excellent examples to be found in the less-developed nations of the world that should be equally recognized, celebrated, and emulated. Therein lies the motivation for this book, *Leadership Matters: An Industrial Engineering Framework for Developing and Sustaining Industry*.

Incidentally, if HFM had originated in the Western world, Mr. Odunayo would be recognized and celebrated along the same lines as Jack Welsh, who was the chairman and chief executive officer (CEO) of General Electric (GE) from 1981 to 2001. Similarly, Tunde's industrial accomplishments would be likened to that of Lee Iacocca (an industrial engineer), the automobile executive at Ford Motor Company in the 1960s and best known for reviving the Chrysler Automobile Corporation as CEO in the 1980s. Iacocca presided over the operations of two of the big three automakers in America. It is of note that Iacocca earned a Bachelor of Science in industrial engineering at Lehigh University in Bethlehem, Pennsylvania. He practiced industrial engineering to the core throughout his career, and it is surmised the core of his success was grounded in this background. What is also of note is Mr. Odunayo did not have any degrees or background in industrial engineering. This should give hope to any reader of similar ilk that they can harness the power of the theories, methods, tools, and techniques published in this book just by reading it. No degree required.

WHY THIS BOOK IS NEEDED

1. To document, profile, and promote templates for good leadership in industry.
2. To capture and preserve the industrial engineering framework of Mr. Tunde Odunayo for future leaders to embrace and utilize with confidence.
3. To make the case that leadership matters and matters of leadership include well-accepted models of industrial engineering such as Triple-C and DEJI Model®.
4. To give hope to leader-readers that shortages in resources such as what HFM experienced (tangible national adversities of non-existent national electricity supply grid, lack of reliable water supply, inadequate transportation infrastructure, and other impediments of industry) can be overcome by committing to operating with a very capable leadership framework.
5. To affirm the fact that leaders who are OGA (One Generation Ahead) of the platform of industry can achieve marvelous successes in building and advancing a new industry, much like Mr. Odunayo did.
6. To show that a captain of industry can exist in a developing country, just as they exist in the developed countries of the world.
7. To demonstrate that great case studies can and do come from less-developed countries in addition to the advanced countries of the Western world.
8. To highlight the importance of sustaining national treasures such as HFM to promote the industrial sector of an entire country, and this should be done by committing to leadership matters.
9. To demonstrate that once the proper leadership anchor is lost in an organization, the organization is doomed to fail.

Authors

Adedeji B. Badiru, PhD, is a Professor of Systems Engineering at the Air Force Institute of Technology (AFIT). He is a registered professional engineer and a fellow of the Institute of Industrial Engineers as well as a Fellow of the Nigerian Academy of Engineering. He earned a BS in industrial engineering, an MS in mathematics, and an MS in industrial engineering at Tennessee University and a PhD in industrial engineering at the University of Central Florida. He is the author of several books and technical journal articles and has received several awards and recognitions for his accomplishments. His special skills, experience, and interests center on research mentoring of faculty and graduate students.

Melinda L. Tourangeau's career began in 1984 as an Air Force ROTC scholarship recipient to Georgia Tech to study lasers to prepare her to work on President Reagan's Star Wars. Her first assignment was at Wright Patterson AFB, Dayton, Ohio, to work in WRDC (now AFRL) and then transition to AFIT. She has worked for four defense industry partners contributing her expertise in program management, electro-optics, and lasers. She is pursuing a PhD in education to study technical leadership in industry. Mrs. Tourangeau is an accomplished technical writer, public speaker, and business leader, devoted to mindful technical leadership for mutual benefit of the US military, industry, and the world.

1 History of the Industrial Engineering Framework

For decades, the discipline of industrial engineering (IE) has provided templates over which industries are developed, advanced, and sustained. The formal definition of IE aptly conveys this assessment.

> Industrial engineering is concerned with the design, installation, and improvement of integrated systems of people, materials, information, equipment, and energy by drawing upon specialized knowledge and skills in the mathematical, physical, and social sciences, together with the principles and methods of engineering analysis and design to specify, predict, and evaluate the results to be obtained from such systems.

The IE framework is well-defined, having been forged by the Industrial Revolution and refined over the last century and a half by technology and the global availability of information through the Internet. Parochially speaking, a framework is a bounded set of principles, methods, tools, techniques, theories, and questions that allows users to operate effectively by adhering to those boundary conditions.

Two tools from the IE framework that have been proven effective are the *Triple-C Principle* (Badiru, 2008) and the *DEJI Systems Model*® (Badiru, 2023). These two tools have proven application track records by managers in the IE field. Enmeshing these two tools into the leadership formula strengthens its potential, making it worthy of emulation by leaders, not only in industrial enterprises but also in other spheres of leadership in business, government, academia, and the military.

Triple-C presents a structured pursuit of goals and objectives through *communication*, *cooperation*, and *coordination*. The complementary DEJI Systems Model presents a structured systems-based process of *design*, *evaluation*, *justification*, and *integration*. Both tools provide guidelines that animates the lesson, "without coordination, there is no success and without integration, there is no progress."

THE SYSTEMS APPROACH TO INDUSTRIAL ENGINEERING

A systems perspective, simply put, is viewing phenomena as inputs, processes, and outputs (Figure 1.1).

A systems framework involves a recognition, appreciation, and integration of all aspects of an organization or a facility. A system is characterized by a collection of interrelated elements working together in synergy to produce a composite output that is greater than the sum of the individual outputs of the components. A systems view of a process facilitates a comprehensive inclusion of all the factors involved in the process. In this regard, we expand the simple systems perspective to a more comprehensive framework, typically known as the ICOM (Input, Constraints, Mechanisms) model as shown in Figure 1.2.

DOI: 10.1201/9781003311348-1

FIGURE 1.1   A simple systems perspective.

Leadership actions permeate every stage of the inputs, processes, and outputs.

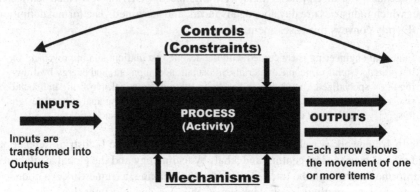

Mechanisms are the physical aspects of the activities
(e.g., people, tools, technology, machines)

Each box represents an activity or sub-process.

Controls constrain and direct activities.
(e.g., quality standards, plans, instructions, directives).

FIGURE 1.2   ICOM model for leadership actions.

LEADERSHIP IN SPORTS TOO

Sports is one area where leadership matters both on and off the field of play. To reaffirm the diversity and versatility of IE, the famed football quarterback, Roger Staubach, during a 2016 interview partially credited the coaching and leadership success of Coach Tom Landry of the Dallas Cowboys to his background in IE and the meticulous planning and attention to detail. Tom Landry received his bachelor's degree in IE from the University of Texas at Austin in 1949. In 1952, he earned a master's degree in IE from the University of Houston.

NOT JUST AN IE TICKET

You don't have to be an industrial engineer to be a great leader, but IE does make it more possible to rise to the rank of great leaders. As was mentioned in the Preamble, Tunde Odunayo was not an industrial engineer, but he led Honeywell Flour Mills very much like an industrial engineer would have done it.

The society often wonders about the following interests (Badiru, 2014):

- How a product can be designed to fit people rather than forcing people to accommodate the product?
- How merchandise layouts can be designed to maximize the profit of a retail store?
- How can hospitals improve patient care while lowering cost?
- How paper companies manage their forests (paper-making raw material) to both increase profits and still ensure long-term availability of trees, while promoting sustainability?
- How the work environment can be designed to enhance comfort and safety while increasing productivity?
- How a fast-food restaurant knows how many and which kinds of burgers to have ready for the lunch-break rush?
- How new car designs can be tested before a prototype is ever built?
- How space exploration can be coordinated to link both management and technical requirements?
- How a military multi-pronged attack can be organized to sustain the supply lines?

Industrial engineers, with a systems-thinking approach, help answer and solve all these operational wonderments. IE thrives on systems perspectives just as systems thrive on IE approaches. One cannot treat IE topics effectively without recognizing a systems perspectives and vice versa. Thus, it makes sense to have a leadership book that is based on the foundational principles of IE.

The formal definition of IE presented in this chapter embodies the various aspects of what an industrial engineer does. The definition is versatile, flexible, and diverse. It can also be seen from the definition that a systems orientation permeates the work of industrial engineers. Some of the major functions of industrial engineers involve the following:

- Design integrated systems of people, technology, process, and methods.
- Develop performance modeling, measurement, and evaluation for systems.
- Develop and maintain quality standards for industry and business.
- Apply production principles to pursue improvements in service organizations.
- Incorporate technology effectively into work processes.
- Develop cost mitigation, avoidance, or containment strategies.
- Improve overall productivity of integrated systems of people, materials, and processes.
- Recognize and incorporate factors affecting performance of a composite system.
- Plan, organize, schedule, and control production and service projects.
- Organize teams to improve efficiency and effectiveness of an organization.
- Install technology to facilitate workflow.
- Enhance information flow to facilitate smooth operations of systems.
- Coordinate materials and equipment for effective systems performance.

An IE colleague, Susan Blake (at Tinker Air Force Base in Oklahoma City), once succinctly defined the profession as "Industrial engineering makes systems function better together with less waste, better quality, and fewer resources."

Leveraging a systems approach is a cornerstone of IE. To remind, a system is a collection of interrelated elements whose collective output is higher than the sum of the individual outputs. So, what is a systems approach? A systems approach, as we define it here, is the recognition that no pursuit should be executed in isolation. Everything interacts with something else, either through interdependencies or through interrelationships. It is through this recognition that everything within the scope of a goal or objective (people, tools, or process) is interconnected to something else. This facilitates a more robust appreciation of what each element brings to the organizational table. Although there may be differing definitions and interpretations of the systems approach, the same basic message is embedded as we have articulated it in the explanation above. Even more importantly, integration, the cornerstone of the DEJI Systems Model (Badiru, 2014), is very essential as the capstone of a systems approach.

IE can be described as the practical application of the combination of engineering fields together with the principles of scientific management and leadership axioms. It is the engineering of work processes and the application of engineering methods, practices, and knowledge to production and service enterprises. IE places a strong emphasis on the understanding of workers and their needs in order to increase and improve production and service activities. IE activities and techniques include the following:

1. Designing jobs (determining the most economical way to perform work).
2. Setting performance standards and benchmarks for quality, quantity, and cost.
3. Designing and installing facilities.

INDUSTRIAL ENGINEERING AND THE INDUSTRIAL REVOLUTION

IE has a proud heritage with a link that can be traced back to the *Industrial Revolution*. Although the practice of IE has been in existence for centuries, the work of Frederick Taylor in the early twentieth century was the first formal emergence of the profession. It has been referred to with different names and connotations. Scientific management was one of the original names used to describe what industrial engineers do. IE principles and frameworks could not have evolved the way it did without the prompting of the legacy of the Industrial Revolution. IE emerged not only to capitalize on innovation but also on the speed of innovation facilitated by the Industrial Revolution. IE harnesses the power of mathematics, social sciences, and many other operational assets to advance the tools, techniques, processes, and practices that best promote industrial production. Leadership fits directly into this organizational treatise.

"*Industry*," the root of the name of IE, clearly explains what the profession is about. The dictionary defines industry generally as the ability to produce and deliver goods and services. The "industry" in IE can be viewed as the application of skills and cleverness to achieve work objectives. This relates to how human effort is harnessed innovatively to carry out work. Thus, any activity can be defined as "industry" because it generates a product, be it service or physical product. A systems view of IE encompasses all the details and aspects necessary for applying skills and cleverness to produce work efficiently. However, the academic curriculum of IE must change, evolve, and adapt to the changing systems environment of the profession.

It is widely recognized that the occupational discipline that has contributed the most to the development of modern society is *engineering*, through its various segments of focus. Engineers design and build infrastructures that sustain the society. These include roads, residential and commercial buildings, bridges, canals, tunnels, communication systems, healthcare facilities, schools, habitats, transportation systems, and factories. Across all of these, the IE process of systems integration facilitates the success of the infrastructures. In this sense, the scope of industrial and systems engineering (ISE) steps through the levels of activity, task, job, project, program, process, system, enterprise, and society. This systems-oriented book presents essential tools to embody this hierarchy of functions in an organization and from a leadership perspective. From the age of horse-drawn carriages and steam engines to the present age of intelligent automobiles and aircraft, the impacts of ISE cannot be mistaken, even though the contributions may not be recognized in the context of the ISE disciplinary identification.

It is essential to recognize the alliance between "industry" and "industrial engineering" as the core basis for the profession. The profession has gone off on too many different tangents over the years. Hence, it has witnessed the emergence of IE professionals who claim sole allegiance to some narrow line of practice, focus, or specialization rather than the core profession itself. Industry is the original basis of IE and should be preserved as the core focus, which should be supported by the different areas of specialization. While it is essential that we extend the tentacles of IE to other domains, it should be realized that over-divergence of practice will not sustain the profession. The continuing fragmentation of IE is a major concern for its unified application in a radically different operating environment of today. A fragmented profession cannot survive for long. The incorporation of systems can help to bind everything together. Notable industrial developments that fall under the purview of the practice of IE range from the invention of the typewriter to the invention of the automobile. Some examples are discussed below.

LEGACY PRODUCTS OF THE INDUSTRIAL REVOLUTION

Writing is a basic means of communicating and preserving records. It is one of the most basic accomplishments of the society. The course of history might have taken a different path if early writing instruments had not been invented at the time that they were. Below is the chronological history of the typewriter:

1714: Henry Mill obtained British patent for a writing machine.
1833: Xavier Progin created a machine that uses separate levers for each letter.
1843: American inventor, Charles Grover Thurber, developed a machine that moves paper horizontally to produce spacing between lines.

1873: E. Remington & Sons of Ilion, NY, manufacturers of rifles and sewing machines, developed a typewriter patented by Carlos Glidden, Samuel W. Soule, and Christopher Latham Sholes, who designed the modern keyboard. This class of typewriters wrote in only uppercase letters but contained most of the characters on the modern machines.

1912: Portable typewriters were first introduced.

1925: Electric typewriters became popular. This made typeface to be more uniform. International Business Machines Corporation (IBM) was a major distributor for this product.

In each case of product development, engineers demonstrate the ability to design, develop, manufacture, implement, and improve integrated systems that include people, materials, information, equipment, energy, and other resources. Thus, product development must include in-depth understanding of appropriate analytical, computational, experimental, implementation, and management processes.

Going further back in history, several developments helped form the foundation for what later became known as IE. In America, George Washington was said to have been fascinated by the design of farm implements on his farm stead in Mt. Vernon. He had an English manufacturer send him a plow built to his specifications that included a mold on which to form new irons when old ones were worn out, or would need repairs. This can be described as one of the early attempts to create a process of achieving a system of interchangeable parts. Thomas Jefferson invented a wooden mold board which, when fastened to a plow, minimized the force required to pull the plow at various working depths. This is an example of early agricultural industry innovation. Jefferson also invented a device that allowed a farmer to seed four rows at a time. In pursuit of higher productivity, he invented a horse-drawn threshing machine that did the work of ten men.

Meanwhile in Europe, the Industrial Revolution was occurring at a rapid pace. Productivity growth, through reductions in manpower, marked the technological innovations of the 1769–1800 Europe. Sir Richard Arkwright developed a practical code of factory discipline. In their foundry, Matthew Boulton and James Watt developed a complete and integrated engineering plant to manufacture steam engines. They developed extensive methods of market research, forecasting, plant location planning, machine layout, work flow, machine operating standards, standardization of product components, worker training, division of labor, work study, and other creative approaches to increase productivity. Charles Babbage, who is credited with the first idea of a computer, documented ideas on scientific methods of managing industry in his book entitled *On the Economy of Machinery and Manufacturers*, which was first published in 1832. The book contains ideas on division of labor, paying less for less important tasks, organization charts, and labor relations. These were all forerunners of modern IE.

Back in America, several efforts emerged to form the future of the IE profession. Eli Whitney used mass production techniques to produce muskets for the U.S. Army. In 1798, Whitney developed the idea of having machines make each musket part so that it could be interchangeable with other similar parts. By 1850, the principle of interchangeable parts was widely adopted. It eventually became the basis for modern mass production for assembly lines. It is believed that Eli Whitney's principle of interchangeable parts contributed significantly to the Union victory during the U.S. Civil War.

Management attempts to improve productivity prior to 1880 did not consider the human element as an intrinsic factor. However, from 1880 through the first quarter of the twentieth century, the works of Frederick W. Taylor, Frank and Lillian Gilbreth, and Henry L. Gantt created a long-lasting impact on productivity growth through consideration of the worker and his or her environment.

Frederick Winslow Taylor (1856–1915) was born in the Germantown section of Philadelphia to a well-to-do family. At the age of 18, he entered the labor force, having abandoned his admission to Harvard University due to an impaired vision. He became an apprentice machinist and pattern-maker in a local machine shop. In 1878, when he was 22, he went to work at the Midvale Steel Works. The economy was in a depressed state at that time. Frederick was employed as a laborer. His superior intellect was very quickly recognized. He was soon advanced to the positions of time clerk, journeyman, lathe operator, gang boss, and foreman of the machine shop. By the age of 31, he was made chief engineer of the company. He attended night school and earned a degree in mechanical engineering in 1883 from Stevens Institute. As a work leader, Taylor faced the following common questions:

"Which is the best way to do this job?"

"What should constitute a day's work?"

These are questions still faced by industrial and systems engineers of today. Taylor set about the task of finding the proper method for doing a given piece of work, instructing the worker in following the method, maintaining standard conditions surrounding the work so that the task could be properly accomplished, and setting a definite time standard and payment of extra wages for doing the task as specified. Taylor later documented his industry management techniques in his book entitled *The Principles of Scientific Management.*

The work of Frank and Lillian Gilbreth coincided with the work of Frederick Taylor. In 1895, on his first day on the job as a bricklayer, Frank Gilbreth noticed that the worker assigned to teach him how to lay brick did his work in three different ways. The bricklayer was insulted when Frank tried to tell him of his work inconsistencies – when training someone on the job, when performing the job himself, and when speeding up. Frank thought it was essential to find one best way to do the work. Many of Frank Gilbreth's ideas were similar to Taylor's. However, Gilbreth outlined procedures for analyzing each step of work flow. Gilbreth made it possible to apply science more precisely in the analysis and design of the work place. Developing *therbligs*, which is Gilbreth spelled backward, as elemental predetermined time units, Frank and Lillian Gilbreth were able to analyze the motions of a worker in performing most factory operations in a maximum of 18 steps. Working as a team, they developed techniques that later became known as work design, methods improvement, work simplification, value engineering, and optimization. Lillian (1878–1972) brought to the engineering profession the concern for human relations. The foundation for establishing the profession of IE was originated by Frederick Taylor and Frank and Lillian Gilbreth.

Henry Gantt's work advanced the management movement from an industrial management perspective. He expanded the scope of managing industrial operations. His concepts emphasized the unique needs of the worker by recommending the following considerations for managing work:

a. Define his task, after a careful study.
b. Teach him how to do it.
c. Provide an incentive in terms of adequate pay or reduced hours.
d. Provide an incentive to surpass it.

Henry Gantt's major contribution is the Gantt chart, which went beyond the works of Frederick Taylor or the Gilbreths. The Gantt chart related every activity in the plant to the factor of time. This was a revolutionary concept for that time. It led to better production planning control and better production control. This involved visualizing the plant as a whole, like one big system made up of interrelated subsystems. Industry has undergone a hierarchical transformation over the past several decades. The drawing below shows how industry has been transformed from one focus level to the next ranging from efficiency of the 1960s to the present-day nano-science trend. IE has progressed from the classical efficiency focus to the present-day digital operations.

In pursuing the applications of ISE, it is essential to make a distinction between the tools, techniques, models, and skills of the profession. *Tools* are the instruments, apparatus, and devices (usually visual or tangible) used for accomplishing an objective. *Techniques* are the means, guides, and processes for utilizing tools for accomplishing the objective. A simple and common example is the technique of using a hammer (a tool) to strike a nail to drive the nail into a wooden work piece (objective). A *model* is a bounded series of steps, principles, or procedures for accomplishing a goal. A model applied to one problem can be replicated and reapplied to other similar problems, provided the boundaries of the model fit the scope of the problem at hand. *Skills* are the human-based processes of using tools, techniques, and models to solve a variety of problems. Very important within the skills set of an industrial engineer are interpersonal skills or soft skills. This human-centric attribute of IE is what sets it apart from other engineering fields.

INDUSTRIAL REVOLUTION AND MODERN LEADERSHIP

Heminger (2014) presents a convincing argument for linking effective modern leadership to the Industrial Revolution. Over the past few decades, a process approach has come to dominate our view of how to conceptualize and organize work. Current approaches to management, such as business process reengineering (BPR) (Hammer and Champy, 1993), Lean (Womack and Jones, 2003), and Six Sigma (Pande et al, 2000), are all based on this concept. Indeed, it seems almost axiomatic today to assume that this is the correct way to understand organizational work. Yet, each of these approaches seems to say different things about processes. What do they have in common that supports using a process approach? And, what do their different approaches tell us about different types of problems with the management of organizational work. To answer these questions, it may help to take a historical look at how work has been done since before the Industrial Revolution up to today.

Prior to the Industrial Revolution, work was done largely by the craftsmen, who underwent a process of becoming skilled in their trade of satisfying customer's wants and needs. Typically, they started as apprentices where they learned the rudiments of their craft from beginning to end, moved on to become journeymen, then craftsmen as they become more knowledgeable, and finally, reaching the pinnacle of their craft as master craftsmen. They grew both in knowledge of their craft and in understanding what their customers wanted. In such an arrangement, organizational complexity was low, with a few journeymen and apprentices working for a master craftsman. But, because work by craftsmen was slow and labor-intensive, only a few of the very wealthiest people could have their needs for goods met. Most people did not have access to the goods that the few at the top of the economic ladder were able to get. There was a long-standing and persistent unmet demand for more goods.

This unmet demand, coupled with a growing technological capability, provided the foundations for the Industrial Revolution. Manufacturers developed what Adam Smith (1776) called the "division of labor," in which complex tasks were broken down into simple tasks, automated where possible, and supervisors/managers were put in place to see that the pieces came together as a finished product. As we moved further into the Industrial Revolution, we continued to increase our productivity and the complexity of our factories. With the huge backlog of unmet demand, there was a willing customer for most of what was made. But, as we did this, an important change was taking place in how we made things. Instead of having a master craftsman in charge who knew both how to make goods as well as what the customers wanted and needed, we had factory supervisors, who learned how to make the various parts of the manufactured goods come together. Attention and focus began to turn inward from the customers to the process of monitoring and supervising complex factory work.

Over time, our factories became larger and ever more complex. More and more management attention needed to be focused inward on the issues of managing this complexity to turn out ever higher quantities of goods. In the early years of the twentieth century, Alfred Sloan at General Motors, did for management what the Industrial Revolution had done for labor. He broke down management into small pieces and assigned authority and responsibility tailored to those pieces. This allowed managers to focus on small segments of the larger organization and to manage according to the authority and responsibility assigned. Through this method, General Motors was able to further advance productivity in the workplace. Drucker (1993) credits this internal focus on improved productivity for the creation of the middle class over the past 100 years. Again, because of the long-standing unmet demand, the operative concept was that if you could make it, you could sell it. The ability to turn out huge quantities of goods culminated in the vast quantities of goods created in the United States during and immediately following World War II. This was added to by manufacturers in other countries which came back on line after having their factories damaged or destroyed by the effects of the war. As they rebuilt and began producing again, they added to the total quantities of goods being produced.

Then, something happened that changed everything. Supply started to outstrip demand. It didn't happen everywhere evenly, either geographically or by industry. But, in ever-increasing occurrences, factories found themselves supplying more than people were demanding. We had reached a tipping point. We went from a world where demand outpaced supply to a world where more and more supply outpaced demand (Hammer and Champy, 1993). Not everything being made was going to sell, at least not for a profit. When supply outstrips demand, customers can choose. And, when customers can choose, they will choose. Suddenly, manufacturers were faced with what Hammer and Champy call the "3 Cs": customers, competition, and change (Hammer and Champy, 1993). Customers were choosing among competing products, in a world of constant technological change. To remain in business, it was now necessary to produce those products that customers will *choose*. This required knowing what customers wanted. But, management and the structure of organizations from the beginning of the Industrial Revolution had been largely focused inward, on raising productivity and making more goods for sale. Managerial structure, information flows, and decision points were largely designed to support the efficient manufacturing of more goods, not on tailoring productivity to the needs of choosy customers.

BUSINESS PROCESS REENGINEERING

A concept was needed that would help organizations focus on their customers and their needs. A process view of work provided a path for refocusing organizational efforts on meeting customer needs and expectations. On one level, a process is simply a series of steps, taken in some order, to achieve some result. Hammer and Champy, however, provided an important distinction in their definition of a process. They defined it as "a collection of activities that takes one or more inputs and creates an output that is of value to the customer" (1993). By adding the customer to the definition, Hammer and Champy provided a focus back on the customer, where it had been prior to the Industrial Revolution. In their 1993 book, *Reengineering the Corporation: A Manifesto for Business Revolution*, Hammer and Champy advocated BPR, which they defined as "the fundamental rethinking and radical redesign of business processes to achieve dramatic improvements in critical, contemporary measures of performance …." In that definition, they identified four words they believed were critical to their understanding of reengineering. Those four words were "fundamental," "radical," "dramatic," and "processes." In the following editions of their book, which came out in 2001 and 2003, they revisited this definition and decided that the key word underlying all of their efforts was the word "process." And, with process defined as "taking inputs, and turning them into outputs of value to a customer," customers and what customers value are the focus of their approach to reengineering.

Hammer and Champy viewed BPR as a means to rethink and redesign organizations to better satisfy their customers. BPR would entail challenging the assumption under which the organization had been operating, and to redesign around their core processes. They viewed the creative use of information technology as an enabler that would allow them to provide the information capabilities necessary to support their processes while minimizing their functional organizational structure.

At roughly the same time that this was being written by Hammer and Champy, Toyota was experiencing increasing success and buyer satisfaction through its use of Lean, which is a process view of work focused on removing waste from the value stream. Womack and Jones (2003) identified the first of the Lean principles as value. And, they state, "Value can only be defined by the ultimate customer." So, once again, we see a management concept that leads organizations back to focus on their customers. Lean is all about identifying waste in a value stream (similar to Hammer and Champy's process) and removing that waste wherever possible. But, the identification of what waste is can only be determined by what contributes or doesn't contribute to value, and value can only be determined by the ultimate customer. So, once again, we have a management approach that refocuses organizational work on the customers and their values.

Lean focuses on five basic concepts: value, the value stream, flow, pull, and perfection. "Value" is determined by the ultimate customer, and the "value stream" can be seen as similar to Hammer and Champy's "process," which focuses on adding value to its customers. "Flow" addresses the passage of items through the value stream, and it strives to maximize the flow of quality production. "Pull" is unique to Lean and is related to the "just-in-time" nature of current manufacturing. It strives to reduce in-process inventory that is often found in large manufacturing operations. "Perfection" is the goal that drives Lean. It is something to be sought after, but never to be achieved. Thus, perfection provides the impetus for constant process improvement.

In statistical modeling of manufacturing processes, sigma refers to the number of defects per given number of items created. Six Sigma refers to a statistical expectation of 3.4 defects per million items. General Electric adopted this concept in the development of the Six Sigma management strategy in 1986. While statistical process control can be at the heart of a Six Sigma program, General Electric and others have broadened its use to include other types of error reduction as well. In essence, Six Sigma is a program focused on reducing errors and defects in an organization. While Six Sigma does not explicitly refer back to the customer for its source of creating quality, it does address the concept of reducing errors and variations in specifications. Specifications can be seen as coming from customer requirements; so again, the customer becomes key to success in a Six Sigma environment.

Six Sigma makes the assertion that quality is achieved through continuous efforts to reduce variation in process outputs. It is based on collecting and analyzing data rather than depending on hunches or guesses as a basis for making decisions. It uses the steps define, measure, analyze, improve, and control (DMAIC) to improve existing processes. To create new processes, it uses the steps define, measure, analyze, design, and verify (DMADV). Unique to this process improvement methodology, Six Sigma uses a series of karate-like levels (yellow belts, green belts, black belts, and master black belts) to rate practitioners of the concepts in organizations. Many companies that use Six Sigma have been satisfied by the improvements that they have achieved. To the extent that output variability is an issue for quality, it appears that Six Sigma can be a useful path for improving quality.

From the above descriptions, it is clear that while each of these approaches uses a process perspective, they address different problem sets and suggest different remedies. BPR addresses the problem of getting a good process for the task at hand. It recognizes that many business processes over the years have been designed with an internal focus, and it uses a focus on the customer as a basis for redesigning processes that explicitly address what customers need and care about. This approach would make sense where organizational processes have become focused on internal management needs, or some other issues, rather than on the needs of the customer.

The Lean methodology came out of the automotive world and is focused on gaining efficiencies in manufacturing. Although it allows for redesigning brand-new processes, its emphasis allows for working with an existing assembly line and finding ways to reduce its inefficiencies. This approach would make sense for organizations that have established processes/value streams with a goal to make those processes/value streams more efficient.

Six Sigma was developed from a perspective of statistical control of industrial processes. At its heart, it focuses on variability in processes and error rates in production and seeks to control and limit variability and errors where possible. It asserts that variability and errors cost a company money, and learning to reduce these will increase profits. Similar to both BPR and Lean, it is dependent on top-level support to make the changes that will provide its benefits. Summarized below are process improvement methods and their respective areas of focus.

- *Business Process Reengineering:* Addresses ineffective, inefficient processes. Focuses on a better process, typically by radical redesign of operations.
- *Lean Methodology*: Addresses waste in the value stream of the organization. Focuses on identifying wasted steps in the value stream and eliminating them where possible.
- *Six Sigma Technique*: Addresses errors and variability of product outputs. Focuses on identifying causes of errors and variable outputs, often using statistical control techniques, and finding ways to control the undesirable variability.

Whichever of these methods is selected to provide a more effective and efficient approach to doing business, it may be important to remember the lessons of the history of work since the beginning of the Industrial Revolution. We started with craftsmen satisfying the needs of a small base of customers. We then learned to increase productivity to satisfy the unmet demand of a much larger customer base, but in organizations that were focused inward on issues of productivity, not outward toward the customers. Now that we have reached a tipping point where supply can overtake demand, we need to again pay attention to customer needs for our organizations to survive and prosper. One of the process views of work may provide the means to do that.

LEADERSHIP DEVELOPMENT REVOLUTION

Khan Academy (Khan Academy, 2022) proposes the incredible change that took place as part of the Industrial Revolution was due to three factors: (a) expanding exchange networks, (b) a marked growing importance of merchant activity and commerce, and (c) the discovery and harvesting of fossil fuels to burn for energy. This was the first time on earth the entire globe forged a network, and its presence formulated a confluence of ideas from myriad cultures and regions. It is argued this was the primary source of innovation. The realization that items could be produced at a rate greater than one could consume them drove the importance of merchant markets. This brought in competition and a second driver for innovation: commerce. In this chapter, we discussed the impact of customer need on production efficiencies. By 1800, the need to sell goods at a lower price and with greater value or quality rose markedly if one were to be successful at market. Incidentally, the sudden appearance of the ease to earn profit and change one's economic status was seductive, to say the least. The frenzy that emerged to focus on selling at market for demand rather than have one's crops seized by dictating aristocrats was too good to pass up. Peasants left the agrarian profession in droves, changing their place and role in society and how and where people lived. Cities grew. Being able to produce more in less time became an obsession. And then, fossil fuels came onto the scene.

The modern IE framework was developed around the turn of the twentieth century as a result of readily available fossil fuels being used to create energy. The rate of production and supply chain logistics accelerated. Planning frameworks and industrial principles needed to be redefined to accommodate the massive rate of change. By the mid-twentieth century, leaders like Jack Welch and Lee Iacocca (Iacocca, 1983) emerged on the scene to capture and leverage the novel advantages that emerged out of the modern industrial framework. They were brilliant in their approaches and left no stone unturned. They found benefit in every pathway and subsystem that was identified as part of modern industrial production. They leveraged the IE framework to maximize outcomes in every area of their business of concern. Therefore, close examination of their approaches and methods, their behaviors, and ultimately their beliefs can render useful scalars for enthusiastic leaders in industry.

Prior to 1700, virtually all production occurred by manpower or horsepower (animal power). The discovery of coal in the mid-eighteenth century in Europe catalyzed a revolution in people, process, and tools. If industrial production had some merits with human or animal power, the merits with fossil fuels outpaced every benefit prior. Now, the race was on. The rate of this change impacted the unsuspecting human by increasing one's daily consumption of energy from 2000 kcals per day in the Paleolithic Era to over 200,000 kcals a day in the early twenty-first century (Khan Academy). People's lives were forever changed.

A mere 120 years ago, the industrialized powers of the earth demonstrated they could wield power and literally form empires by taking under-industrialized regions under their command and control. With today's modern exploitation of information (commonly called the Information Age), it is unclear who or what will have power. If the same information is available to everyone around the world, what will be the deciding factors for progress and prosperity? One can argue the continued progress and ability to produce goods at a rapid, efficient rate to make them available to world markets will grant an edge to those countries that can collect that wealth. And while the governments of those countries will necessarily reap some of that reward, the companies that produced those goods will assuredly benefit. While leadership can be looked at through the eyes of a single individual, it's clear the impact of leadership is far greater than that, and in fact in today's world, it is global. Every advantage, or every method or practice available for aspiring leaders, should be examined and exposed. Given the inarguable success of modern IE principles to pro-liferate the success of production, it makes sense to apply those same principles to the one who's put in charge of seeing that production succeed. Now, 120 years later, we are seeing a complete reexamination of leadership and its impact on society will be felt for many years to come. Leadership models and practices must adapt accordingly.

Organizational structure is the foundation of any industry. The Bible speaks to the house built on sand perishing in a storm, but the house built on rock will stand up to the storm. A similar case can be made for putting extreme emphasis and early excessive attention on the *organizational structure*.

It's pretty clear the best industrial engineers need two things: knowledge of how systems operate and acumen to improve tasks and processes. If one were to pro-mote the industrial engineer to leader, the new leader would need two more skills: trust and respect. The combination of these four elements will wield the most effec-tive leadership for industry. Since organizational structure is the most important foundation for any industry, an effective industry leader would possess both keen knowledge of how an organizational structure operates and decisive acumen to make recommendations for efficacious improvement of organizational structures. He or / she would also command trust and respect. In the coming pages, it will be shown how Tunde accomplished this admirably, as did the Western hemisphere examples, Jack Welch and Lee Iacocca.

When one thinks of the Industrial Revolution, it is easy to reflect only the impact on industrial production practices. Perhaps the reader also now recognizes that era of humanity also dawned the formal study of IE. IE principles and frameworks could not or would not have evolved without the Industrial Revolution, and IE likely emerged not only to capitalize on innovation but also on the speed of the innovation and the rate of change of innovation. IE harnesses the power of mathematics, social sciences, and a host of other factors to expose tools, techniques, processes, and prac-tices that best promote industrial production. Leadership is arguably attempting to do the same thing.

The most important component of leadership is, therefore, a blend of three commitments:

a. People as human beings.
b. Their acting cooperatively to realize outcomes that are most efficient and optimal.
c. Growth and/or profit, and not at the expense of people.

IE intends to bring about the most optimal and efficient processes, tools, and tech-niques to render successful production outcomes. When IE principles are applied to leadership, comparable optimization of personnel management can be achieved.

Change is one thing; the speed of change and the rate of change is entirely another. Much like the first and second derivatives of an equation, one can learn a great deal about change when it is studied from the perspective of how quickly things change and how fast the speed of change changed. People have a hard enough time coping with change; then, when change happens faster and faster, there is even more unrest. An industrial leader would do well to keep this in mind and manage its impact on employees.

Tunde Odunayo did this through his commitment to employee training, education, and general career development along the path of gradual and incremental programs. Under his leadership, Honeywell Flour Mills (HFM) particularly sent selected employees for job-themed technical training programs. Employees so trained are placed in jobs and tasks that directly leverage their newly acquired skills. This has the positive impact of job satisfaction, career motivation, and organizational morale advancement.

When industrial production proliferated, people started going to work, not working from home. This meant leadership had to take on a new responsibility and a new ability – the ability to manage large groups of people and organize them to produce goods. The Industrial Revolution spawned the industrial leader. Industrial production paved the way for IE, defined as the interplay between mathematics, statistics, social sciences, psychology, economics, management (Sassani, 2017), and information and computer technology. The IE framework exposes tools and techniques, such as Triple-C and DEJI® Model, to assist those working in industry to perform better. Industrial leaders know this and are on the lookout for best practices. Tunde knew this and made finding best practices an easter egg hunt. Jack Welch and Lee Iacocca also demonstrated a keen ability to find what works best and harness it. See Gerstner (2002) for a related case example.

Every leader wants to excel. In some cases, it may be to turn around a flailing company and in other cases, it may be to maintain the success of a company. When a leader "succeeds" a predecessor, if the company is doing well, it behooves the leader to examine what they were doing well, i.e., study their methods of "success." Perhaps it isn't coincidence that these two words share the same root. This is a helpful play on words for burgeoning leaders when they take over a company like Honeywell Flour Mills and wish to stave off a drastic turn and ultimate failure.

In the next chapter, we will draw upon the background presented here to propose a formula for leadership success.

REFERENCES

Badiru, Adedeji B., editor (2014) Handbook of Industrial & Systems Engineering, Second Edition. Taylor & Francis CRC Press, Boca Raton, FL.

Drucker, Peter F. (1993) The Post Capitalist Society. Harper Collins, New York.

Gerstner, Louis (2002) Who Says Elephants Can't Dance? Harper Collins, New York.

Hammer, Michael and Champy, James A. (1993, 2001, 2003) Reengineering the Corporation: A Manifesto for Business Revolution. Harper Business, New York.

Heminger, Alan (2014) "Industrial Revolution, Customers, and Process Improvement," Chapter in Handbook of Industrial & Systems Engineering, Second Edition, Adedeji Badiru, editor. Taylor & Francis CRC Press, Boca Raton, FL.

Iacocca, Lee Anthony (1983) The Rescue and Resuscitation of Chrysler. *The Journal of Business Strategy*, 4(1), 67. https://go.openathens.net/redirector/liberty.edu?url=https://www.proquest.com/scholarly-journals/rescue-resuscitation-chrysler/docview/1295087644/se-2

Khan Academy (2022) *World History Project – 1750 to the Present; Unit 3: Industrialization*. https://www.khanacademy.org/humanities/whp-1750/xcabef9ed3fc7da7b:unit-3-industrialization

Pande, Peter S., Neuman, Robert S. and Cavanaugh, Roland R. (2000) The Six Sigma Way: How GE, Motorola, and Other Top Companies Are Honing Their Performance. McGraw-Hill, New York.

Sassani, Farrokh. (2017) Industrial Engineering Foundations - Bridging the Gap between Engineering and Management. Mercury Learning and Information, Herndon, VA.

Smith, Adam (orig. 1776, 2012) The Wealth of Nations. Simon & Brown, New York, NY.

Womack, James P. and Jones, Daniel T. (2003) Lean Thinking: Banish Waste and Create Wealth in Your Corporation. Free Press, New York.

2 Industrial Engineering Framework for Leadership

Good outcome is predicated on good leadership.

Adedeji Badiru

A PRACTICAL FORMULA FOR LEADERSHIP

If you had a formula for success, would you use it over and over? From the background that was presented in the last chapter, it is now appropriate to propose a formula that is based on an unusual approach to leadership: the systems perspective and the industrial engineering (IE) framework.

To draw an analogy for the systems perspective to leadership, a leader observes the environment around him or her, assimilates that information, cogitates, and processes it (by thinking and making decisions), and then takes action. Even for the staunchest non-engineer reading this book, it is hopefully reliably clear that leadership can be modeled as a system.

The other piece of the formula is the IE framework. By presenting useful tools, techniques, models, and methods from the IE framework, leaders can enhance their approach and decisions to realize their leadership goals with confidence and valuable effect. This book will formulate the systems perspective and the components of the IE framework into a usable formula that can be used over and over for success.

One more item to buttress this new leadership formula involves the real-life situations. Case studies paint a vibrant picture of success (or failure). In this book, we will be showcasing the impressive success of Mr. Tunde Odunayo (Badiru, 2015), the unassuming president of the Honeywell Flour Mills in Nigeria, who exacted incredible success in his leadership term, only to have it fall apart when he stepped down. Both sides of this case study, as well as other leaders, will be showcased throughout this book to demonstrate the viability of a leadership formula and in what context to apply it. We are now in a position to propose our formula for leadership, as shown in Figure 2.1.

Figure 2.2 presents the pictorial representation of the theme and premise of this book. The figure shows how and where the techniques of Triple-C and DEJI Systems Model align with the tenets of IE. This pictorial framework is presented and referred to again and again through the sections of this book to ensure that the book stays on the core message of leadership improvement through a systems perspective.

DOI: 10.1201/9781003311348-2

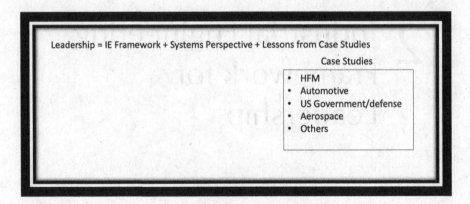

FIGURE 2.1 A practical formula for leadership.

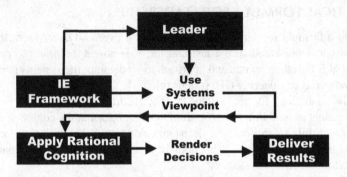

FIGURE 2.2 Comprehensive leadership environment.

Leadership swims in the seas of many functions, as depicted in Figure 2.3. Thus, leadership must be versatile and adept at many things.

An interesting leadership treatise is presented by John A. White, one of the foremost leaders of IE, who, in his book entitled *Why It Matters: Reflections on Practical Leadership* (White, 2022), highlighted many of the leadership topics advocated in this book. In White's book, the following required attributes were defined for good leadership.

Leaders must:

1. Be advocates for the ideal
2. Lead from the power of their ideas and who they are
3. Make work purposeful
4. Be courageous
5. Build efficacy in others
6. Be gracious under pressure
7. Protect and direct the culture
8. Demonstrate composure under pressure
9. Demonstrate positivity whenever negativity arises

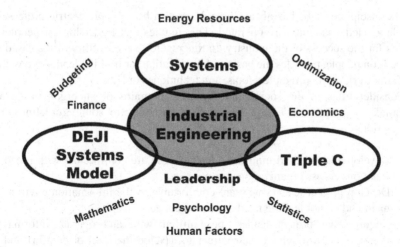

FIGURE 2.3 IE framework for leadership.

10. Accept criticism gracefully
11. Market their core values
12. Have a greater sense of purpose than the moment
13. Value truth more than loyalty
14. Value mercy more than justice

Out take on this list is that, in a society where justice is paramount many people will claim an exception to the 14th "must." At the same time, advocates of the recent movements on diversity, equity, inclusion, and justice may place a high value on justice as a concomitant asset of mercy rather than an opposite.

ISSUES IN LEADERSHIP DEVELOPMENT

The basic lesson of leadership via the tools and techniques of IE can best be summarized in the following quote:

Think like an industrial engineer, act like an industrial engineer.

Adedeji Badiru's motto for the practice of industrial engineering

There are environmental factors that impinge upon how leadership is manifested in an organization. We consider the famous satirical quote by Petronius Arbiter (210 B.C.), a Roman courtier during the reign of Nero, who is reported to have written "We trained hard, but it seemed that every time we were beginning to form into teams we would be reorganized. I was to learn later in life that we tend to meet any new situation by reorganizing; and what a wonderful method it can be for creating the illusion of progress while producing confusion, inefficiency and demoralization." What Arbiter noted reflects what we often see in organizational settings of the past, presently, and expectedly in the future (see Tokar, 2020)

The discipline of IE has always been known to be a people-centric profession. People are led, machines are operated. This implies that leadership (of people) is crucial for the success of the industry. In this regard, the versatility of IE, based on its focus on people, provides the basis and justification for framing leadership within the framework of the principles, tools, and techniques of IE.

Considerations in the social and cognitive domains of an organization are essential for proper leadership development. Below are some guidelines and expectations:

- Develop adaptive and innovative leaders who are comfortable with broad functions across organizational units.
- Develop project leaders who are comfortable in the information environment and do not feel the need to micromanage.
- Require that project leaders be familiar with and operate information systems throughout their career to develop the level of comfort and knowledge required to leverage the most from the available information systems. This requires formal education, organizational training, and continuing education programs.
- Develop project leaders who are comfortable operating in a continuous, concurrent planning process versus the continuous, sequential planning process. This requires a type of "benevolent hierarchy," where leaders operating at varying levels and subordinate personnel are freely sharing information and working more in a collegial systems-type environment to accomplish project goals.

If the benefits of systems structure are to be realized, a comprehensive change is needed across the operating environment in terms of recognizing and responding to all the requirements of personnel interfaces, technology, resource allocation, and work process. First, change must occur in the physical and information domains. New communications, organizational structure (OS), and other supporting infrastructure and information flow must be developed. Information age technologies developed in specific application domains must be adapted for the current project environment. Second, change must occur in the social and cognitive domains. This entails changing the ways people, organizations, and their processes interact. It requires new ways of selecting, training, and assigning project personnel, and it depends on the ability to overcome traditional obstacles to information sharing and functional collaboration. It requires that trust and confidence be fostered amongst a variety of people from diverse backgrounds (functions and services). Breaking down existing stovepipes becomes an imperative within any project for systems that limit information shareability between functions. Breaking down barriers that limit collaboration and interoperability between one department and another is equally essential. The specific organization structure that is selected and utilized within a leadership framework can facilitate success with the challenges cited earlier. By doing these things, a latent benefit will emerge: the organization will organically eliminate waste.

PREPARING THE PERSONNEL

For effective management practices, the organization's pursuits are best framed as a collection of projects. Thus, project management techniques are required for leading an organization. One such proven technique is categorizing workers based on their skills and proficiency and then assigning them to jobs and tasks commensurate with project requirements. The organization should use its personnel management systems to codify the new skill sets required for multidimensional projects and then workers should be allocated to those projects in the same personnel management system.

When the personnel management system is used effectively, it bypasses stove-piped assignments in favor of cross-training and collaborative execution of tasks.

Cultural change is required to break down information sharing barriers and create an environment that better encourages resource sharing. Project personnel at all levels must be quicker and more flexible in order to provide an adequate response to the rapid evolution of the project scope based on the speed of operations and the improvement in shared situational awareness. A hierarchical and directive model will not always work. The goal should be a professional and collegial atmosphere that emphasizes a rapid and collaborative interchange of information rather than a reactive response to project requirements.

Training is at the core of every successful operation. Success of the project system is almost guaranteed with, or, is reflected by direct and realistic training of personnel in the specific requirements of each and every project. A systems approach requires familiarity with and confidence in the project team itself as well as the available tools, tactics, techniques, policies, procedures, and work process. Most projects are information rich. The challenge is how to extract and leverage the inherent information to accomplish a project.

LEADERSHIP OF THE PROJECT OFFICE

A Project Office is the anchor for the successful execution of a project. Below is a list of desired characteristics of a project office, from a leadership perspective:

- Establish and operate a project management office (PMO) to serve as the command post for project operations.
- The PMO must develop a plan to train and integrate project personnel across the project system.
- Operating a project system requires flexibility and must support a degree of ad hoc responsiveness to quickly adjust processes and structural design to maximize the sharing of information and increase collaboration in varying situations.
- The PMO can be the command-and-control center with on-the-move capabilities from the management level down to the secretarial support staff.

WORK DISTRIBUTION FOR LEADERSHIP EFFECTIVENESS

There are three major categories of breakdown structures. The *project breakdown structure (PBS)* is a project-specific structure that refers to the breakdown of the tasks to be performed in a specific project. This is often referred to as the work breakdown structure (WBS). The *OS* refers to the organizational structure with which a project is to be carried out. The *organizational breakdown structure (OBS)* refers to the identification and organization of the resources required to carry out the activities associated with a project. OBS is a model that provides a way of organizing resources into groups for better management within an organization, which may be a company, a project, or a division of a large enterprise. OBS can be used to keep track of resource allocation and specific work assignments. There is a strong interdependency between OBS and WBS. A good breakdown of work helps in estimating resource requirements more accurately. Good team organization is essential for the success of OBS. For example, a mixed team comprised of individuals with different technical backgrounds is required for technology-based projects. OBS team requirements are:

- Background of the leader: experience, education, technical knowledge
- Scope of work to be accomplished: hours, skills, locations
- Estimate of the number and type of personnel required
- Equipment and work space requirements
- Reporting procedures

SELECTING AN ORGANIZATIONAL STRUCTURE

After project planning, the next step is the selection of the OS for the project. This involves the selection of an OS that shows the management line and responsibilities of the project personnel. Any of several approaches may be utilized. Before selecting an OS, the project team should assess the nature of the job to be performed and its requirements. The structure may be defined in terms of functional specializations, departmental proximity, standard management boundaries, operational relationships, or product requirements.

Large and complex projects should be based on well-designed structures that permit effective information and decision processes. The primary function of an organizational design is to facilitate effective information flow. Traditional organization models consist of the decision-making, bureaucracy, social, and systems structures. The decision-making structure handles the policies and general directions of the overall organization. The bureaucracy is concerned with the administrative processes. Some of the administrative functions may not be directly relevant to the main goal of the organization, but they are, nonetheless, deemed necessary. Bureaucratic processes are potential sources of delay for public projects because of government involvement. The social structure facilitates amiable interactions among the personnel. Such interpersonal relationships are essential for the group effort needed to achieve company objectives. The systems structure can better be described as the link among the various synergistic segments of the organization.

Many organizations still use the traditional or classical organization structure. A traditional organization is often utilized in service-oriented companies. The structure is sometimes referred to as the pure functional structure because groups with similar functional responsibilities are clustered at the same level of the structure. The positive characteristics of the structure are:

Availability of broad personnel base
Identifiable technical line of control
Grouping of specialists to share technical knowledge
Collective line of responsibility
Possible assignment of personnel to several different projects
Clear hierarchy for supervision
Continuity and consistency of functional disciplines
Possibility for departmental policies, procedures, and missions

However, the traditional structure does have some negative characteristics:

No one individual is directly responsible for the total project.
Project-oriented planning may be impeded.
There may not be a clear line of reporting up from the lower levels.
Coordination is complex.
A higher level of cooperation is required between adjacent levels.
The strongest functional group may wrongfully claim project authority.

FORMAL AND INFORMAL ORGANIZATIONAL STRUCTURES

The formal organization structure represents the officially sanctioned structure of a functional area. The informal organization, on the other hand, develops when people organize themselves in an unofficial way to accomplish an objective that is in line with the overall project goals. The informal organization is often very subtle in that not everyone in the organization is aware of its existence. Both formal and informal organizations are practiced in every project environment. Even organizations with strict hierarchical structures, such as the military, still have some elements of informal organization.

LEADERSHIP SPAN OF CONTROL

The functional organization calls attention to the form and span of management that are suitable for company goals. The span of management (also known as the span of control) can be wide or narrow. In a narrow span, the functional relationships are streamlined with fewer subordinate units reporting to a single manager. The wide span of management permits several subordinate units to report to the same boss. The span of control required for a project is influenced by a combination of the following:

Level of planning required
Level of communication desired
Effectiveness of delegating authority
Dynamism and nature of the subordinate's job
Competence of the subordinate in performing his or her job

Given a project environment that is conducive, the wide span of management can be very effective. From a motivational point of view, workers tend to have a better identification with upper management since there are fewer hierarchical steps to go through to reach the top. More professional growth is possible because workers assume more responsibilities. In addition, the wide span of management is more economical because of the absence of extra layers of supervision. However, the narrow span of management does have its own appeal in situations where there are several mutually exclusive skill levels in the organization.

FUNCTIONAL ORGANIZATIONAL STRUCTURE

The most common type of formal organization is known as the functional organization, whereby people are organized into groups dedicated to particular functions. Depending on the size and the type of auxiliary activities involved in a project, several minor, but supporting, functional units can be developed for a project.

Projects that are organized along functional lines are normally resident in a specific department or area of specialization. The project home office or headquarters is located in the specific functional department. For example, projects that involve manufacturing operations may be under the control of the vice president of manufacturing while a project involving new technology may be assigned to the vice president for advanced systems. The advantages of a functional organization structure are:

Improved accountability
Discernible line of control
Flexibility in manpower utilization
Enhanced comradeship of technical staff
Improved productivity of specially skilled personnel
Potential for staff advancement along functional path
Use of home office can serve as a refuge for project problems

The disadvantages of a functional organization structure are:

Divided attention between project goals and regular functions
Conflict between project objectives and regular functions
Poor coordination of similar project responsibilities
Unreceptive attitude by the surrogate department
Multiple layers of management
Lack of concentrated effort

It is difficult to separate the project environment from the traditional functional environment. There must be an integration. The DEJI Systems Model (Badiru, 2023), discussed later on in this book, offers an effective framework for accomplishing systems integration. The project management approach affects the functional management approach and vice versa. Since tasks are the basic components of a project and tasks are the major focus of functional endeavors, they form the basis for the integration of the project and functional environments.

PRODUCT ORGANIZATION STRUCTURE

Another approach to organizing a project is to use the end product or goal of the project as the determining factor for personnel structure. This is often referred to as the pure project organization or, simply, project organization. The project is set up as a unique entity within the parent organization. It has its own dedicated technical staff and administration. It is linked to the rest of the system through progress reports, organizational policies, procedures, and funding. The interface between product-organized projects and other elements of the organization may be strict or liberal depending on the organization.

The product organization is common in large project-oriented organizations or organizations that have multiple product lines. Unlike the functional structure, the product organization decentralizes functions. It creates a unit consisting of specialized skills around a given project or product. Sometimes referred to as team, task force, or product group, the product organization is common in public, research, and manufacturing organizations where specially organized and designated groups are assigned specific functions. A major advantage of the product organization is that it gives the project members a feeling of dedication to and identification with a particular goal.

A possible shortcoming of the product organization is the requirement that the product group be sufficiently funded to be able to stand alone without sharing resources or personnel with other functional groups or programs. The product group may be viewed as an ad hoc unit that is formed for the purpose of a specific goal. The personnel involved in the project are dedicated to the particular mission at hand. At the conclusion of the mission, they may be re-assigned to other projects. The product organization can facilitate the most diverse and flexible grouping of project participants and permits highly dedicated attention to the project at hand. The advantages of the product organization structure are:

Simplicity of structure
Unity of project purpose
Localization of project failures
Condensed and focused communication lines
Full authority in the project manager
Quicker decisions due to centralized authority
Skill development due to project specialization
Improved motivation, commitment, and concentration
Flexibility in determining time, cost, and performance trade-offs
Accountability of project team to one boss (project manager)
Individual acquisition and maintenance of expertise on a given project

The disadvantages are:

Narrow view of project personnel (as opposed to global organization view)
Mutually exclusive allocation of resources (one man to one project)
Duplication of efforts on different but similar projects

Monopoly of organizational resources
Concern about life after the project
Reduced skill diversification

MATRIX ORGANIZATION STRUCTURE

The matrix organization is a popular choice of management professionals. A matrix organization exists where there are multiple managerial accountability and responsibility for a job function. It attempts to combine the advantages of the traditional structure and the product organization structure. In pure product organization, technology utilization and resource sharing are limited because there is no single group responsible for overall project planning. In the traditional organization structure, time and schedule efficiency are sacrificed. Matrix organization can be defined as follows:

> Matrix organization is a structure of management that facilitates maximum resource utilization and increased performance within time, cost, and performance constraints.

There are usually two chains of command: horizontal and vertical. The horizontal line deals with the functional line of responsibility while the vertical line deals with the project line of responsibility. The project manager has total responsibility and accountability for project success. The functional managers have the responsibility to achieve and maintain high technical performance of the project.

The project that is organized under a matrix structure may relate to specific problems, marketing issues, product quality improvement, and so on. The project line in the matrix is usually of a temporary nature while the functional line is more permanent. The matrix organization is dynamic. Its actual structure is determined by the prevailing activities of the project.

The matrix organization has the following advantages:

Good team interaction
Consolidation of objectives
Multilateral flow of information
Lateral mobility for job advancement
Opportunity to work on a variety of projects
Efficient sharing and utilization of resources
Reduced project cost due to sharing of personnel
Continuity of functions after project completion
Stimulating interactions with other functional teams
Cooperation of functional lines to support project efforts
Home office for personnel after project completion
Equal availability of company knowledge base to all projects

The disadvantages of matrix organization are:

Slow matrix response time for fast-paced projects
Independent operation of each project organization

High overhead cost due to additional lines of command
Potential conflict of project priorities
Problems of multiple bosses
Complexity of the structure

Despite its disadvantages, the matrix organization is widely used in practice. Its numerous advantages seem to outweigh the disadvantages. In addition, the problems of the matrix organization structure can be overcome with good project planning, which can set the tone for a smooth organization structure. Matrix organization is a collaborative effort between product and functional organization structures. It permits both vertical and horizontal flows of information. The matrix model is sometimes called a multiple-boss organization. It is a model that is becoming increasingly popular as the need for information sharing increases. For example, large technical projects require the integration of specialties from different functional areas. Under matrix organization, projects are permitted to share critical resources as well as management expertise. The following project situations are suitable for implementing a matrix organization:

1. When the primary outputs of an organization are numerous, complex, and resource critical.
2. When a complicated design calls for innovation and widespread expertise.
3. When expensive, sophisticated, and scarce technologies are needed in designing, building, and testing products.
4. When emergency response and flexibility are required for a project.

Traditionally, industrial projects are conducted in serial functional implementations such as R&D, engineering, manufacturing, and marketing. At each stage, unique specifications and work patterns may be used without consulting the preceding and succeeding phases. The consequence is that the end product may not possess the original intended characteristics. For example, the first project in the series might involve the production of one component while the subsequent projects might involve the production of other components. The composite product may not achieve the desired performance because the components were not designed and produced from a unified point of view. In today's interdependent market-oriented projects, such lack of a unified design will lead to overall project failure.

The major appeal of matrix organization is that it attempts to provide synergy within groups in an organization. This synergy can be realized if certain ground rules are observed when implementing a matrix organization. Some of those rules are as follows:

1. There must be an individual who is devoted full time to the project.
2. There must be both horizontal and vertical channels for communication.
3. There must be a quick access and conflict resolution strategy between managers.
4. All managers must have an input into the planning process.
5. Functional and project managers must be willing to negotiate and commit resources.

MIXED ORGANIZATION STRUCTURE

Another approach for organizing a project is to adopt a combined implementation of the functional, product, and matrix structures. This permits the different structures to coexist simultaneously in the same project. In an industrial project, for example, the project of designing a new product may be organized using a matrix structure while the subproject of designing the production line may be organized along functional lines. The mixed model facilitates flexibility in meeting special problem situations. The structure can adapt to the prevailing needs of the project or the needs of the overall organization. However, a disadvantage is the difficulty in identifying the lines of responsibility within a given project. There is a wide array of mixed organizations based on the matrix and other structures. These range from a single product/project manager with strong dependence on the functional organization, to a large product/project organization with little dependence on the functional organization. The functional personnel may be located in the PMO or in a separate geographical location. They may be fully dedicated to a single project manager or may serve many project managers. In the next section, we present new ideas for unique organization structures that cater to specific project situations.

ALTERNATE ORGANIZATION STRUCTURES

In addition to the traditional OS, new structures are often needed to address unique organizational or project needs. Some of the structures presented hereafter will, no doubt, be of a temporary nature and new ones will always be needed to accommodate unique needs that develop within the project system.

BUBBLE ORGANIZATION STRUCTURE

The bubble organization introduced by Badiru (2019), also called the *blob organization*, is a structure that allows functional teams to rally around a central project goal. This may be suitable for grass-roots movement among society groups canvassing for a national need. The bubble structure will, most often, be temporary in nature. It disorganizes as soon as its goal is accomplished or it is deemed no longer worthwhile.

MARKET ORGANIZATION STRUCTURE

As world markets expand, it is important to be more responsive to the changing market forms. The market organization structure permits a project to adapt to market conditions. The defunct Honeywell Flour Mills, whose case study is presented Chapter 5, followed the market organizational structure in its heyday. The approach to OS is particularly adept in evolving markets in developing countries.

CHRONOLOGICAL ORGANIZATION STRUCTURE

The chronological organization is suitable for projects where time sequence is very essential in organizing tasks. This is for time-critical sets of tasks. A training program is suitable for the use of a chronological organization structure.

SEQUENTIAL ORGANIZATION STRUCTURE

The sequential organization is similar to the chronological structure except that magnitude or quality of output rather than time is the basis for organizing the project. The quality of the output at each stage of the organization structure is needed to carry out the functions of the organization sequentially. A value-added production facility is an example of a system that is suitable for a sequential organization structure.

MILITARY ORGANIZATION STRUCTURE

The military organization (Badiru, 2019) follows a strict hierarchical structure and chain of command. It discourages informal lines of communication or responsibility. The name does not necessarily refer to the traditional military structure or configuration, but as in the military command structure, the block at the top of the military organization structure is notably more powerful than the lower blocks.

POLITICAL ORGANIZATION STRUCTURE

A political organization structure (Badiru, 2019) can be viewed as a rotary type of structure that is dynamic with respect to a time cycle. It has a very large base. The large base is collectively more powerful than the few blocks at the top. This may also be referred to as a democratic organization structure.

AUTOCRATIC ORGANIZATION STRUCTURE

An autocratic organization may be viewed as the reverse of the political organization structure. There is a single block at the top that is infinitely more powerful than the rest of the organization. Despite the large number of blocks at the lower levels of the structure, the top block remains almightily powerful. A major difference between the military and the autocratic organization structures is that there is a higher prospect that the block at the top of the military structure can be replaced. An autocratic organization structure for industry sustainment does not work well in democratic societies.

LEADERSHIP OF PROJECT TRANSFER

Very much like going through a technology transfer, an organization can also experience a "project transfer." Project and organizational transfers are important aspects of organization management. The leadership and OS that are in effect during a project can influence the transfer of the final product of a project. The organic (internal) OS and the external OS must be linked by a discernible transfer path, which may consist of transfers of products, ideas, concepts, hardware, software, personnel, and/or decisions. The transfer path shows how products, ideas, concepts, and decisions move from one project environment to another. The receiving organization (referred to as the *transfer target*) uses the transferred elements to generate new products, ideas, concepts, and decisions, which follow a reverse transfer path to the *transfer source*. Thereby, both project environments operate on a symbiotic basis, with each contributing something to the other. Project transfer can be achieved in various forms as outlined in the following text.

1. **Transfer of complete project products.** In this case, a fully developed product is transferred from one project to another project. Very little product development effort is carried out at this receiving point. However, information about the operations of the product is fed back to the source so that necessary product enhancements can be pursued. Hence, the recipient generates product information which serves as a resource for future work of the transfer source.

2. **Transfer of procedures and guidelines.** In this transfer mode, procedures and guidelines are transferred from one project to another. The project blueprints are implemented locally to generate the desired services and products. The use of local raw materials and personnel is encouraged for the local operations. Under this mode, the implementation of the transferred procedures can generate new operating procedures that can be fed back to enhance the original project. With this symbiotic arrangement, a loop system is created whereby both the transferring and the receiving organizations obtain useful benefits.

3. **Transfer of project concepts, theories, and ideas.** This strategy involves the transfer of the basic concepts, theories, and ideas associated with a project. The transferred elements can then be enhanced, modified, or customized within local constraints to generate new project outputs. The local modifications and enhancements have the potential to initiate an identical project; a new related project; or a new set of project concepts, theories, and ideas. These derived products may then be transferred back to the transfer source.

The important questions to ask about project transfer include the following:

What exactly is being transferred?
Who is receiving the transferred elements?
What is the cost of the transfer?
How is this project similar to previous projects?
How is this project different from previous projects?
Are the goals of the projects similar?
What is expected from the transferred project?
Is in-house skill adequate to use the transferred project?
Is the prevailing management culture receptive to the new project?
Is the current infrastructure capable of supporting the project?
What modifications to the project will be necessary?

The selection of an appropriate transfer mode is particularly important for projects that cross national boundaries as discussed in the next section.

LEADING AND ORGANIZING MULTINATIONAL ORGANIZATIONS

Projects and organizations that cross national boundaries either in concept or in implementation have unique characteristics that create project management problems. In multinational projects, individual organizational policies are not enough to govern operations. Factors that normally influence these projects include:

- Territorial laws and regulations
- Geographical segregation and restricted access
- Time differences
- Different scientific standards of measure
- Trade agreements
- Different government and political ideologies
- Different social, cultural, and labor practices
- Different stages of industrialization
- National security concerns
- Protection of proprietary technology information
- Strategic military implications
- Traditional national allies and adversaries
- Taxes, duties, and other import/export charges
- Foreign currency exchange rates
- National extradition/protection agreements
- Paperwork, permits, and restrictions
- Health, weather, and environmental considerations
- Poor, slow, or incompatible communication links
- Different native languages

International communication is perhaps one of the most difficult to deal with among all the factors listed previously. The task of international transfer of technology and mutual project support takes on critical dimensions because of differences in the structures, objectives, and interests of the different countries involved, not to mention the time differences. One common communication problem is that information destined for another country may have to pass through several levels of approval before reaching the point of use. The information is subject to all types of distortions and perils in its arduous journey. The integrity of the information may not be preserved as it is passed from one point to another. When implementing international projects, the following considerations should be reviewed:

1. Product
   a. Type of product expected
   b. Portability of the product
   c. Product maintenance
   d. Required training
   e. Availability of spare parts
   f. Feasibility for intended use
   g. Local versus overseas productions

2. Technology
   a. Local availability of required technology
   b. Import/Export restrictions on the technology
   c. Implementation requirements
   d. Supporting technologies
   e. Adaptation to local situations
   f. Lag between development and applications
   g. Operational approvals required
3. Political and social environment
   a. Leadership and consistency of national policies
   b. Political and social stability
   c. Management views
   d. Cultural adaptations
   e. Bureaucracies
   f. Structures of formal and informal organizations
   g. Acceptance of foreigners and formation of relationships
   h. Immigration laws
   i. Ethnicity
   j. General economic situation
   k. Religious situations
   l. Population pressures
   m. Local development plans
   n. Decision-making bureaucracies
4. Labor
   a. Union regulations
   b. Wage structures
   c. Personnel dedication, loyalty, and motivation
   d. Educational background and opportunities
   e. Previous experience
   f. Management relationships
   g. Economic condition and level of contentment
   h. Interests, attitudes, personalities, and leisure activities
   i. Taxation policies
   j. Logistics of employee relocation
   k. Productivity consciousness
   l. Demarcation of private and business activities
   m. Local communication practices and facilities
   n. Local customs
   o. Language barriers
5. Market
   a. Market needs
   b. Inflation
   c. Stability
   d. Variety and availability of products
   e. Cash, credit, and billing requirements
   f. Exchange rates

      g. National budget and gross national product
      h. Competition and size of market
      i. Transportation facilities
6. Plant and residential amenities
      a. Location
      b. Structural condition
      c. Accessibility
      d. Facilities available
      e. Proximity to business centers
      f. Topography
      g. Basic amenities (water, light, sewage, etc.)
7. Financial services
      a. Banking
      b. International money transfer
      c. Currency strength and stability
      d. Local sources of capital
      e. Interest rates
      f. Efficiency in conducting transactions
      g. Investment laws

Because of all these various factors, international project managers must undergo more extensive training than their conventional counterparts. A foreign manager in an international project must be open-minded, flexible, adaptive, and able to learn quickly. Since he or she will be working in an unfamiliar combination of social, cultural, political, and religious settings, he or she must have a keen sense of awareness and should be unassuming and responsive to local practices. An evaluation of the factors presented earlier in the context of the specific countries involved should help the international project manager to be better prepared for his or her expanded role.

INTERCONNECTED VERSUS INTERDEPENDENT LEADERSHIP

Words matter when describing the attributes and characteristics of leadership of an organization. For example, we distinguish between interconnected leadership as opposed to interdependent leadership. Interconnected leadership represents operational and leadership connectivity among the top leaders within an organization or across collaborating organizations. By contrast, interdependent leadership refers to necessary interfaces among mutually dependent leaders. This may be the type of relationship between an organization and its suppliers.

TRANSACTIONAL LEADERSHIP VERSUS STRATEGIC LEADERSHIP

Transactional leadership is demonstrated by a leader who is well versed in getting things done on the spur of the moment as needs develop. This leader is able to quickly gather his or her subordinates to coordinate actions to get things done. By comparison, strategic leadership is more focused on long-range pursuit of opportunities, even those not yet fully defined. This leader uses intuition, foresight, and/or forecasting to lean forward into the future to make decisions that affect the organization, not only within the present scenario but also to position the organization to embrace and leverage future opportunities. Mr. Tunde Odunayo of the Honeywell Flour Mills case study demonstrated his leadership by using both transactional and strategic leadership qualities. That was the crux of his success as a leader. Yes, leadership matters.

SUMMARY OF SYSTEMS-BASED LEADERSHIP FOR SUSTAINMENT

Based on the preceding contents of this chapter, we now present a closing discussion of systems-based leadership sustainment and the importance of systems thinking in leadership of business and industry, where people are involved.

Frameworks sustain things. As pieces and parts move, ebb and flow, frameworks lockdown the necessary components for continued success of operation. Frameworks (the good ones) are not easy to identify. Much like the architecture for a building is not self-evident, a good framework for leadership doesn't just appear. Good frameworks take time to develop. Incidentally, most frameworks appear out of need a way to fill an identified gap. A gap is usually made evident by failure, especially where success was the predecessor.

Leadership failure is not new. In fact, it spans the dawn of time. Moses never entered the promised land; Jesus was crucified. Rome fell; Hitler committed suicide. All that is remains firmly in the hands of leadership, yet leadership is not treated as an essential element of the organizational system, or maybe it is superfluously, and so every time a leader changes, the system underneath also changes, and, sometimes, not for the better. When we recognize that a change in leadership is a bona fide change to a system, we can better accept that treating leadership as a system, in itself, is a prudent approach.

Since leadership theory isn't doing the trick, maybe we need another approach. Engineering has promise. After all, $E = mc^2$, right? The sum of all forces at a node must equal 0, yes? $P = IV$, $V = IR$, etc., always (unless we are traveling at the speed of light). Since we're not, let's consider the probability of success of applying engineering principles to leadership success.

The basis of engineering principles is if you apply the laws (equations) correctly, the result is predictable, given gravity and sub light speed. Entirely predictable. With certainty. One need only identify initial conditions (correctly) and the relationships, that is, the equations. What if leadership were approached with such certainty? Or predictability? Would we celebrate or at least be optimistic that the outcomes were desirable? When an equation reveals a result we don't desire, we change the initial conditions until the equation yields the result we want. Or we change the equation.

If the laws of leadership were broken down essentially to equations what would happen? So, here's the rub – leaders are human, and the systems they lead are human. The equation contains variables that have infinite possible variables. But it also has constants and predictable variables. If we can find a way to write the equations so they reflect the moving parts and predict results, then we've done it.

An IE framework for leadership demands no less and it can be done. How? By examining great leaders and their behaviors, distilling them down to common factors and then assembling them into a framework. We are now back where we started.

Southern New Hampshire University defines organizational leadership as follows:

> Organizational leadership is a management approach in which leaders help set strategic goals for the organization while motivating individuals within the group to successfully carry out assignments in service to those goals.

The important items to extract from this definition are "strategic goals," and "carry out assignments." The two-pronged model for successful leadership, transactional and strategic, is fully represented in this definition. If a leader is going to run an organization effectively, he or she should constantly be considering both sides of this coin. If the leader is going to expect constituents to complete tasks based on strategic goals, then they themselves should possess the acumen to be an organizational leader. In other words, organizations should be run by organizational leaders, and organizational leaders must be both transactional and strategic. The success of organizations requires both these scalars, so the leader better possess them.

An IE framework for leadership flies in the face of classical management theory. An IE framework necessitates there are components working together, feeding back, controlling, and that any time the system can be changed by changing external or internal conditions. Classical management theory says there is one best way; use the right tools and set people on the right path, and things will get done. There is also an element of treating people more like resources than human beings, seeing them as tools. This, too, diverges from the IE framework. The human element of IE principles recognizes that people are not structured, not predictable, and they are likely the most potent source of change in the system. An effective IE leader will see this instinctively. When they approach any problem, or any venture, seeing the people as human beings properly sets the system's initial conditions to an appropriate level of expectation and prediction.

It might seem unconventional to apply IE principles to leadership. Leadership has been carved up into so many different theories that began with traits (early twentieth century), evolved to behaviors (mid-twentieth century), and then incorporated various social factors to construct modern theories such as transformational, transactional (not the one being offered in this book), servant, and laissez-fare. Those conventional theories possess principles and frameworks that are primarily linear – if an employee does this, then do this; if the organization is going through this, then do that. Or the leader will choose to do, "that." IE principles suggest leadership is a system, too, possessing all the parts of a (engineering) system, including feedback loops, control loops, open to the environment (i.e., not in a lab), and ultimately, there is no real control over anything. Furthermore, all systems are prone to change. A systems approach means you prepare for that change when possible, and recognize it is hardly ever static. Leadership theory is grandly idealistic at the end of the day. An IE framework will ground leadership in reality, and what one can really expect from exercising leadership.

This brings up the subject of entropy, classically defined as natural decay of a system leading to disorder. Or having the *tendency* to decay. Even the best (most perfectly) defined system is subject to this outcome. Any prudent approach to leadership should necessarily consider that system as being prone to decay also. Tunde did well to step down when he did. From the outside, he went out on top. In his heart he knew his luck had run out and decay was about to set in whether he planned against it or not. This natural time element needs to be faced with dignity and respect. And while one might hope the swan song is self-evident, it is more of a visceral knowing than an empirical examination of the evidence. This moment in time will be left to the reader. When you exercise this option, write to us (the authors) and tell us how you knew it was time to go.

Another laudable aspect of systems theory that will be graceful when applied to leadership is in designing goals that are adaptable. There is no one best way. The sooner a leader recognizes this, the better. Allowing for myriad ways to achieve an outcome makes room for people's propensity for disarray and demonstrates a level of trust in people to get it done for the mission's sake. This goes a long way in engendering the ultimate trust and respect that galvanizes employees to follow a leader wherever he or she may lead.

Systems are viewed as the whole being greater than the sum of the parts. To consider components of leadership separately is helpful to bolster the theory and even develop andragogy, but to enter the real world and not see that all the theory in the world will not explain the magic and the enigma that is great leadership is folly. This makes it prudent to treat leadership as a system. When one is outlining the design of a system, one often draws bubbles with a question mark to display a component that cannot be explained, but one knows it has a bearing on system performance. Let's call this Mojo – the hidden ingredient. Defining leadership with a system framework leaves room to design in these hidden elements. Through this case study, those hidden elements will be revealed.

Organizations are in a dynamic, interdependent, interconnected relationship with the environment. Leadership is, too. By its very nature, the human occupying the leadership seat is clearly dynamic (moody, prone to emotions, health issues, hungry), and eternally dependent on and connected to the people being led. Those people are also subject to environmental factors, placing the leader only one level away from whatever set off his or her employee that morning. If a leader isn't careful, those environmental factors will weigh on them, too. The subparts are always interrelated.

Organizational components or factors are traditionally considered to be technological, organizational, and environmental. What does the organizational component mean? Parochially, it is the people-process-tools that make achieving goals possible. There are inputs (resources, information, action) that are subject to processing and productizing (becoming work products) to render outputs (products, decisions, solutions, plans, schedules, goals). When examining leadership through this lens, the individual behaves much like the organization is behaving, taking inputs and processing them into outputs.

Systems are characterized by dynamic relationships. Moving parts, or, the nature of the connections between subsystems is subject to change. Much like the staircases in Harry Potter's Hogwarts School, they change often unexpectedly and seemingly illogically. One has to almost step back to figure out where this change will lead. Nothing could be more prudent for a leader. Just like Harry and his companions didn't just jump on the staircase as it started to move, they stood back and observed where it would land and adjudicated its impact to their destination: Gryffindor's dormitories. Leaders must step back often, assessing impact and deciding how to act. The change could be a one-time event, or it could mean the entire system is going to function this way from now on. That is a prudent observation to make, which brings this discussion back to leadership. A leader, at any time, can and should change direction, sometimes illogically (or seemingly so) to render an outcome. Once trust is established, these changes can be brought in handily, with little resistance. Tunde knew this truism and he exercised it by the way he managed the strategic affairs of Honeywell Flour Mills, as shown later in Chapter 5.

All systems experience feedback. This is one limitation of classical leadership theory in that it really doesn't provide for the impact of feedback on the framework. Classic leadership theory cannot account for all the possible permutations that should be left to models and best practices, hence this book. The important thing to note is that feedback is inherently present in all systems, and leadership is not spared. All systems benefit from observing feedback loops, exploiting them, and leveraging them for greater outputs. In the next chapter, we will demonstrate how feedback has been incorporated into the leadership framework: through the DEJI Systems Model.

REFERENCES

Badiru, Adedeji B. (2015) Folaranmi Babatunde Odunayo – Festschrift in Honour of a Business Leader: Silent Works, Great Works. ABICS Publications, Dayton, OH.

Badiru, Adedeji B. (2019) Project Management: Systems, Principles, and Applications, Second Edition. Taylor & Francis CRC Press, Boca Raton, FL.

Badiru, Adedeji B. (2023) Systems Engineering Using DEJI Systems Model: Design, Evaluation, Justification, and Integration with Case Studies and Applications. Taylor & Francis CRC Press, Boca Raton, FL.

Tokar, Sofia (2020, January 16). *What Is Organizational Leadership?* snhu.edu. Retrieved October 29, 2022, from https://www.snhu.edu/about-us/newsroom/business/what-is-organizational-leadership

White, John A. (2022) Why It Matters: Reflections on Practical Leadership. Greenleaf Book Group Press, Austin, TX.

3 The DEJI Systems Model® for Leadership

LEADERSHIP AND CHANGE

> If you want things to stay as they are, things will have to change.
>
> *−Giuseppe Tomasi di Lampedusa*

A good leader is someone who embraces and facilitates change, from a human systems perspective. The human side of industrial engineering is what sets the discipline apart from other engineering disciplines (Badiru, 2023a). In this regard, from a systems perspective, soft skills form the bastion of the practice of industrial engineering. Hard skills form the quantitative basis for leading and managing the workforce. Using a systems framework, industrial engineers enmesh qualitative and quantitative tools and techniques to manage integrated systems of people, tools, and process. The premise of this chapter is to combine the systems viewpoint and the human cognitive reasoning to improve functional efficiency and productivity in the work environment. A key systems tool discussed in the chapter is the DEJI Systems Model®, which provides a structural pathway for human-based work design, work evaluation, work justification, and work integration (Badiru, 2019). The quote below demonstrates how soft skills facilitate building, actuating, and managing teams.

> Working together productively requires that the work be designed appropriately to permit teamwork.
>
> *−Adedeji Badiru*

DESIGN OF WORK

The planning, organizing, and coordination of work elements all fall under a category of design defined by the DEJI Model. It is essential that a structured approach be applied to work design right at the outset. Retrofitting a work element after problems develop not only impedes the overall progress of work in an organization but also leads to an inefficient use of limited human and material resources. This stage of the model is expected to guide work designers onto the path of strategic thinking about work elements down the line rather than just the tactical manipulation of work for the present needs. In this regard, Badiru (2016) says, "Right next to innovation, structured methods for producing effective work results are a survival imperative for every large organization."

Under the methodology of DEJI Systems Model, normal work planning would fall in the category of design. That is, work planning is work design, which must be handled with strategic thinking, as explained in the next paragraph.

DOI: 10.1201/9781003311348-3

Work is the foundation of everything we do. Having some knowledge is not enough. The knowledge must be applied to do something in the pursuit of objectives. Work management facilitates the application of knowledge and willingness to actually accomplish tasks. Where there is knowledge, willingness often follows. But it is work execution that actually gets things done. From the very basic tasks to the very complex endeavors, work management must be applied to get things done. It is, thus, essential that systems thinking be a part of the core of every work pursuit in business, industry, education, government, and even at home. In this regard, a systems approach is of utmost importance because work accomplishment is a "team sport" that has several underlying factors as elements of the overall work system. Thus, DEJI Systems Model offers a way to achieve a well-designed work plan. After the work design is completed, the next stage is the evaluation of work.

EVALUATION OF WORK

Following the design of a work element, the DEJI Model calls for a formal evaluation of the intended purpose of the work vis-à-vis other work elements going on in the organization. Such an evaluation may lead to a need to go back and re-design the work element. Evaluation can be done as a combination of both qualitative and quantitative assessment of the work element, depending on the specific nature of the work, the main business of the organization, and the managerial capabilities of the organization.

JUSTIFICATION OF WORK

According to the concept of the DEJI Model, not only should a work element be designed and evaluated but it should also be formally and rigorously justified. If this is not done, errant work elements will creep into the organizational pursuits. What is worth doing is worth doing well. Otherwise, it should not be done at all. The principles of Lean operations (Agustiady and Badiru, 2012) suggest weeding out functions that do not add value to the organizational goal. In this regard, each and every work element needs to be justified. But it should be realized that not all work elements are expected to generate physical products in the work environment. A work element may be justified on the basis of adding value to the wellbeing of the worker with respect to mental, emotional, spiritual, and physical characteristics. The point of this stage of the DEJI Model is to ensure that the work element is needed at all. Or, the do-nothing alternative is always an option.

INTEGRATION OF WORK

This last stage of the DEJI Model is of utmost importance, but it is often neglected. The model affirms that the most sustainable work elements are those that fit within the normal flow of operations, existing practices, or other expectations within an organization. Does the work fit in? Will a new work element under consideration be an extraneous pursuit or a detraction in the overall work plan? If a work element is not integrated with other normal pursuits, it cannot be sustained for the long haul. This is why many organizations suffer from repeated program starts and stops. For example, in as much as worker wellness programs are desirable pursuits in an organization, they cannot be sustained if not integrated into the culture and practices of the organization. Unintegrated flash-in-the-plan programs, activities, and work elements often fall by the wayside over time. For this stage of the DEJI Model, work elements must be tied to the end goal of the person and the organization. Badiru (2016) remarks that "aimless work is so insidious because it tends to covertly masquerade as fruitful labor," which is not connected to real organizational goals. This is the premise of applying the multi-dimensional hierarchy of needs (of the worker and the organization) as discussed in a subsequent section of this chapter.

Work is the means to accomplishing a goal. For the purpose of the theme of this book, work is literally considered as an activity to which strength, mental acuity, and resources are applied to get something done. This may involve a sustained physical and/or mental effort to overcome impediments in the pursuit of an outcome, an objective, a result, or a product. In an operational context, work can be viewed as a process of performing a defined task or activity, such as research, development, operations, maintenance, repair, assembly, production, administration, sales, software development, inspection, data collection, data analysis, teaching, and so on. The opening quote in this chapter signifies the meaning of work as a means to an end goal, where workers thrive and the work effort succeeds. If you understand your work from a system's viewpoint, you will enjoy the work and want to do more of it. In this book, we view work as a "work system" rather than work in isolation or disconnected from other human endeavors.

A systems view of work is essential because of several factors and diverse people that may be involved in the performance of the work. There are systems and subsystems involved in each execution of work whether small or large, whether simple or complex, and whether localized or multi-locational. For an activity to be workable consistently, all the attendant factors and issues must be taken into account. If some crucial factors are neglected, the *workability* of the activity may be in jeopardy. For the purpose of a systems view of work, we define a system as a collection of interrelated elements (subsystems) working together synergistically to generate a collective and composite outcome (value) that is higher than the sum of the individual outcomes of the subsystems. Even simple tasks run the risk of failure if some minute subsystem is not accounted for. The following two specific systems engineering models are used for the purpose of the theme of this book:

1. The V-Model of Systems Engineering
2. The DEJI Model of Systems Quality Integration (Badiru, 2023b)

Figure 3.1 presents an illustration of the V-Model applied to a manufacturing enterprise consisting of a series of work elements. Although the model is most often used in software development process, it is also applicable to hardware development as well as general work in systems engineering. In the model, instead of moving down in a linear fashion, the process steps are bent upward after the coding phase, to form the typical V shape. The V-Model demonstrates the relationships between each phase of the development life cycle and its associated phase of testing. The horizontal and vertical axes represent time or project completeness (left-to-right) and the level of abstraction, respectively.

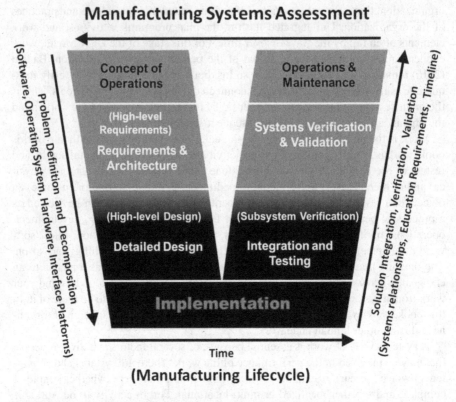

FIGURE 3.1   V-Model of systems engineering for manufacturing.

There are several different ways that a work system can be developed and delivered using the V-model. The best development strategy depends on how much the work analyst knows about the system for which work is being designed. Three basic design strategies mentioned below can be used.

ONCE-THROUGH APPROACH

In this case, we plan, specify, and implement the complete work system in one pass through the V shape. This approach, also sometimes called the "waterfall" approach, works well if the vision is clear, the requirements are well understood and stable, and there is sufficient funding. The problem is that there isn't a lot of flexibility or opportunity for recovery if the vision, work environment, or the requirements change substantially.

INCREMENTAL APPROACH

Here, we plan and specify the work system and then implement it in a series of well-defined increments or phases, where increment delivers a portion of the desired end goal. This is like moving through value-adding increments of the work. In this case, we are making one pass through the first part of the V shape and then iterating through the latter part for each phased-in increment. This is a common strategy for field equipment deployment where system requirements and design can be incrementally implemented and deployed across a given area in several phases and several projects.

EVOLUTIONARY APPROACH

In this approach, we plan, specify, and implement an initial system capability, learn from the experience with the initial system, and then define the next iteration to address issues and extend capabilities (or add value). Thus, we refine the concept of operations, add and change system requirements, and revise the design as necessary. We will continue with successive iterative refinements until the work system is complete. This strategy can be shown as a series of "Vs" that is placed end-to-end since system operation on the right side of the "V" influences the next iteration. This strategy provides the most flexibility but also requires project management expertise and vigilance to make sure the development stays on track. It also requires patience from the stakeholders as the design moves along in incremental stages.

Figure 3.2 presents an illustration of the application of the DEJI Model for systems integration of work factors. The key benefit of the DEJI Model is that it moves the effort systematically through the stages of design, evaluation, justification, and

FIGURE 3.2 Framework for the application the DEJI Model to systems integration.

integration. The approach pings the work analyst about what needs to be addressed in each stage so that the ball is not dropped on critical requirements. The greatest aspect of the DEJI Model is the final stage that calls out the need to integrate the work with other efforts within the work environment. If there is a disconnect, then the work may end up being a misplaced effort. If a work effort is properly integrated, then it will be sustainable.

SYSTEMS DEFINITION FOR SOFT SKILLS

A system is defined as a collection of interrelated elements working together synergistically to achieve a set of objectives. Any work is essentially a collection of interrelated activities, tasks, people, tools, resources, processes, and other assets brought together in the pursuit of a common goal. The goal may be in terms of generating a physical product, providing a service, or achieving a specific result. This makes it possible to view any work as a system that is amenable to all the classical and modern concepts of systems management.

TECHNICAL SYSTEMS CONTROL

Classical technical system control focuses on control of the dynamics of mechanical objects, such as a pump, electrical motor, turbine, rotating wheel, and so on. The mathematical basis for such control systems can be adapted (albeit in iconical formats) for management systems, including work management. This is because both technical and managerial systems are characterized by inputs, variables, processing, control, feedback, and output. This is represented graphically by input-process-output relationship block diagrams. Mathematically, it can be represented as:

$$z = f(x) + \varepsilon,$$

where

z = output
$f(.)$ = functional relationship
ε = error component (noise, disturbance, etc.)

For multi-variable cases, the mathematical expression is represented as vector-matrix functions as shown below:

$$Z = f(x) + E$$

where Z is the output, $f(x)$ is the functional relationship, and E is the error. Regardless of the level or form of mathematics used, all systems exhibit the same input-process-output characteristics, either quantitatively or qualitatively. The premise of this book is that there should be a cohesive coupling of quantitative and qualitative approaches in managing a work system. In fact, it is this unique blending of approaches that makes systems application for work management more robust than what one will find in mechanical control systems, where the focus is primarily on quantitative representations.

SOFT SKILLS AND ORGANIZATIONAL PERFORMANCE

Systems engineering efficiency and effectiveness are of interest across the spectrum of work management for the purpose of improving organizational performance. Managers, supervisors, and analysts should be interested in having systems engineering serve as the umbrella for improving work efforts throughout the organization. This will get everyone properly connected with the prevailing organizational goals as well as create collaborative avenues among the personnel. Systems application applies across the spectrum of any organization and encompasses the following elements:

- Technological systems (e.g., engineering control systems and mechanical systems)
- Organizational systems (e.g., work process design and operating structures)
- Human systems (e.g., interpersonal relationships and human-machine interfaces)

A systems view of the world makes everything work better and work efforts more likely to succeed. A systems view provides a disciplined process for the design, development, and execution of work both in technical and non-technical organizations. One of the major advantages of a systems approach is the win-win benefit for everyone. A systems view also allows full involvement of all stakeholders and constituents of a work center. This is very well articulated by the Chinese saying shown below:

Tell me and I forget;
Show me and I remember;
Involve me and I understand.
–Confucius, Chinese
Philosopher

For example, the pursuit of organizational or enterprise transformation is best achieved through the involvement of everyone, from a systems perspective. Every work environment is very complex because of the diversity of factors involved, including the following:

- The worker's overall health and general wellbeing
- The worker's physical attributes
- The worker's mental abilities
- The worker's emotional stability
- The worker's spiritual interests
- The worker's psychological profile

There are differing human personalities. There are differing technical requirements. There are differing expectations. There are differing environmental factors. Each specific context and prevailing circumstance determine the specific flavor of what can and cannot be done in the work environment. The best approach for effective work management is to adapt to what each work's requirements and specifications. This requires taking a systems view of the work. The work systems approach presented in this book is needed for "working across" organizations, countries, across cultures, and across unique nuances of each project. This is an essential requirement in today's globalized and intertwined personal and professional goals. A systems view requires a disciplined embrace of multidisciplinary execution of work in a way that each component complements other components in the work system. Formal work management represents an excellent platform for the implementation of a system approach. A comprehensive work management program requires control techniques, such as operations research, operations management, forecasting, quality control, and simulation to achieve goals. Traditional approaches to management use these techniques in a disjointed fashion, thus ignoring the potential interplay among the techniques. The need for an integrated systems-based work management worldwide has been recognized for decades. As long ago as 1993, the World Bank reported that a lack of systems accountability led to several worldwide project failures. The bank, which has loaned more than $300 billion to developing countries over the last half century, acknowledged that there has been a dramatic rise in the number of failed projects around the world. In other words, the work efforts failed. A lack of an integrated system approach to managing the projects was cited as one of the major causes of failure.

More recent reports by other organizations point to the same flaws in managing global projects and to the need to apply better project management to major projects. Press headlines in April 2008 highlighted that "Defense needs better management of projects." This was in the wake of government audit that reveals gross inefficiencies in managing large defense projects. In a national news released on April 1, 2008, it was reported that auditors at the Government Accountability Office (GAO) issued a scathing review of dozens of the Pentagon's biggest weapons systems, citing that "ships, aircraft, and satellites are billions of dollars over budget and years behind schedule."

According to the review, "95 major systems have exceeded their original budgets by a total of $295 billion; and are delivered almost two years late on average." Further, "none of the systems that the GAO looked at had met all of the standards for best management practices during their development stages."

Among programs noted for increased development costs were the "Joint Strike Fighter and Future Combat Systems." The costs of those programs had risen "36 percent and 40 percent, respectively," while C-130 avionics modernization costs had risen 323%. And, while Defense Department officials have tried to improve the procurement process, the GAO added that "significant policy changes have not yet translated into best practices on individual programs." In the view of this book, a failed program is an indicator of failed work efforts. A summary of the report of the accounting office reads:

> Every dollar spent inefficiently in developing and procuring weapon systems is less money available for many other internal and external budget priorities, such as the

global war on terror and growing entitlement programs. These inefficiencies also often result in the delivery of less capability than initially planned, either in the form of fewer quantities or delayed delivery to the warfighter.

In as much as the military represents the geo-political-economic landscape of a nation, the above assessment is representative of what every organization faces, whether public or private. In systems-based project management, it is essential that related techniques be employed in an integrated fashion so as to maximize the total project output. One definition of systems project management (Badiru, 2012) is stated as follows:

> Systems project management is the process of using systems approach to manage, allocate, and time resources to achieve systems-wide goals in an efficient and expeditious manner.

The above definition calls for a systematic integration of technology, human resources, and work process design to achieve goals and objectives. There should be a balance in the synergistic integration of humans and technology. There should not be an over-reliance on technology, nor should there be an over-dependence on human processes. Similarly, there should not be too much emphasis on analytical models to the detriment of common-sense human-based decisions.

SYSTEMS FRAMEWORK FOR LEADERSHIP

Systems engineering is growing in appeal as an avenue to achieve organizational goals and improve operational effectiveness and efficiency. Researchers and practitioners in business, industry, and government are all embracing systems engineering implementations. So, what is systems engineering? Several definitions exist. Below is one quite comprehensive definition:

> Systems engineering is the application of engineering to solutions of a multi-faceted problem through a systematic collection and integration of parts of the problem with respect to the lifecycle of the problem. It is the branch of engineering concerned with the development, implementation, and use of large or complex systems. It focuses on specific goals of a system considering the specifications, prevailing constraints, expected services, possible behaviors, and structure of the system. It also involves a consideration of the activities required to assure that the system's performance matches the stated goals. Systems engineering addresses the integration of tools, people, and processes required to achieve a cost-effective and timely operation of the system.

INCOSE (International Council on Systems Engineering) defines systems engineering as follows:

> Systems Engineering is an interdisciplinary approach and means to enable the realization of successful systems. It focuses on defining customer needs and required functionality early in the development cycle, documenting requirements, then proceeding with design synthesis and system validation while considering the complete problem.

Systems engineering integrates all the disciplines and specialty groups into a team effort forming a structured development process that proceeds from concept to production to operation. Systems engineering considers both the business and the technical needs of all involved with the organizational goals.

LEADING HUMAN WORK SYSTEMS

Logistics can be defined as the planning and implementation of a complex task, the planning and control of the flow of goods and materials through an organization or manufacturing process, or the planning and organization of the movement of personnel, equipment, and supplies. Complex projects represent a hierarchical system of operations. Thus, we can view a project system as collection of interrelated projects all serving a common end goal. Consequently, we present the following universal definition:

> Work systems logistics is the planning, implementation, movement, scheduling, and control of people, equipment, goods, materials, and supplies across the interfacing boundaries of several related projects.

Conventional organizational management must be modified and expanded to address the unique logistics of work systems.

ANALYZING SYSTEMS CONSTRAINTS

Systems management is the pursuit of organizational goals within the constraints of time, cost, and quality expectations. The iron triangle model depicted in Figure 3.3 shows that project accomplishments are constrained by the boundaries of quality, time, and cost. In this case, quality represents the composite collection of project requirements. In a situation where precise optimization is not possible, there will have to be trade-offs between these three factors of success. The concept of iron triangle is that a rigid triangle of constraints encases the project. Everything must be accomplished within the boundaries of time, cost, and quality. If better quality is expected, a compromise along the axes of time and cost must be executed, thereby altering the shape of the triangle. The trade-off relationships are not linear and must be visualized in a multi-dimensional context. This is better articulated by a 3-D view of the systems constraints as shown in Figure 3.4. Scope requirements determine the project boundary and trade-offs must be done within that boundary. If we label the eight corners of the box as (a), (b), (c) …, (h), we can iteratively assess the best operating point for the project. For example, we can address the following two operational questions:

1. From the point of view of the project sponsor, which corner is the most desired operating point in terms of combination of requirements, time, and cost?
2. From the point of view of the project executor, which corner is the most desired operating point in terms of combination of requirements, time, and cost?

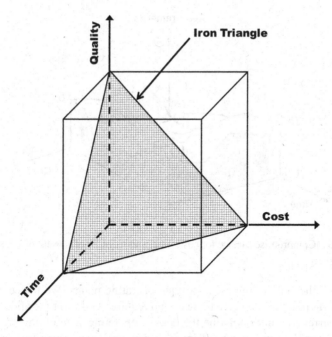

FIGURE 3.3   Systems constraints of cost, time, and quality within iron triangle.

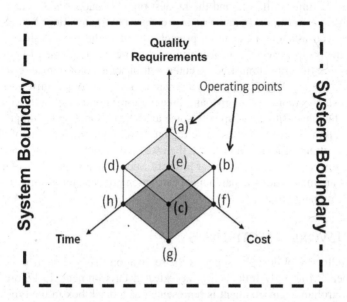

FIGURE 3.4   Systems constraints of cost, time, and quality within iron triangle.

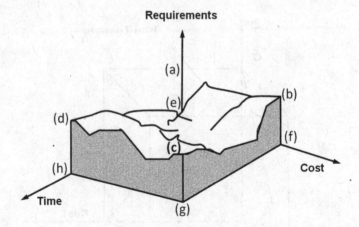

FIGURE 3.5   Compromise surface for cost, time, and requirements trade-off.

Note that all the corners represent extreme operating points. We notice that point (e) is the do-nothing state, where there are no requirements, no time allocation, and no cost incurrence. This cannot be the desired operating state of any organization that seeks to remain productive. Point (a) represents an extreme case of meeting all requirements with no investment of time or cost allocation. This is an unrealistic extreme in any practical environment. It represents a case of getting something for nothing. Yet, it is the most desired operating point for the project sponsor. By comparison, point (c) provides the maximum possible case for requirements, cost, and time. In other words, the highest levels of requirements can be met if the maximum possible time is allowed and the highest possible budget is allocated. This is an unrealistic expectation in any resource-conscious organization. You cannot get everything you ask for to execute a project. Yet, it is the most desired operating point for the project executor. Considering the two extreme points of (a) and (c), it is obvious that the project must be executed within some compromise region within the scope boundary. Figure 3.5 shows a possible view of a compromise surface with peaks and valleys representing give-and-take trade-off points within the constrained box. The challenge is to come up with some analytical modeling technique to guide decision-making over the compromise region. If we could collect sets of data over several repetitions of identical projects, then we could model a decision surface that can guide future executions of similar projects. Such typical repetitions of an identical project are most readily apparent in construction projects, for example residential home development projects.

HUMAN WORK EFFECTIVENESS

Systems influence philosophy suggests the realization that you control the internal environment while only influencing the external environment. In Figure 3.6, the inside (controllable) environment is represented as a black box in the typical input-process-output relationship. The outside (uncontrollable) environment is bounded

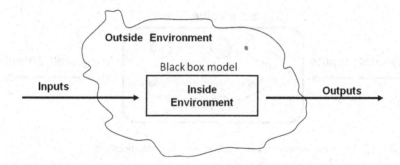

FIGURE 3.6   Outside versus inside environments of a project.

by the cloud representation. In the comprehensive systems structure, inputs come from the global environment are moderated by the immediate outside environment, and are delivered to the inside environment. In an unstructured inside environment, work functions occur as blobs as illustrated in Figure 3.7. A "blobby" environment is characterized by intractable activities where everyone is busy, but without a cohesive structure of input-output relationships. In such a case, the following disadvantages may be present:

- Lack of traceability
- Lack of process control
- Higher operating cost
- Inefficient personnel interfaces
- Unrealized technology potentials

Organizations often inadvertently fall into the blobs structure because it is simple, low-cost, and less time-consuming, until a problem develops. A desired alternative is to model the project system using a systems value-stream structure as shown in Figure 3.8. This uses a proactive and problem-preempting approach to execute projects. This alternative has the following advantages:

- Problem diagnosis is easier
- Accountability is higher
- Operating waste is minimized
- Conflict resolution is faster
- Value points are traceable

FIGURE 3.7   Blobs of work in unstructured programs.

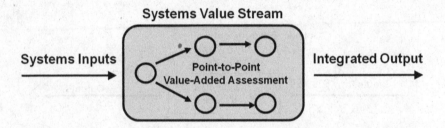

FIGURE 3.8   Systems value-stream structure for work enhancement.

LEADERSHIP ASSESSMENT OF HUMAN WORK

A technique that can be used to assess overall value-added components of a process improvement program is the systems value model (SVM), which is an adaptation of the manufacturing system value (MSV) model presented by Troxler and Blank (1989). The model provides an analytical decision aid for comparing process alternatives. Value is represented as a p-dimensional vector:

$$V = f\left(A_1, A_2, \ldots, A_p\right)$$

where $A = \left(A_1, \ldots, A_n\right)$ is a vector of quantitative measures of tangible and intangible attributes. Examples of process attributes are quality, throughput, capability, productivity, cost, and schedule. Attributes are considered to be a combined function of factors, x_1, expressed as:

$$A_k\left(x_1, x_2, \ldots, x_{m_k}\right) = \sum_{i=1}^{m_k} f_i\left(x_i\right)$$

where $\{x_i\}$ = set of m factors associated with attribute $A_k\left(k = 1, 2, \ldots, p\right)$ and f_i = contribution function of factor x_i to attribute A_k. Examples of factors include reliability, flexibility, user acceptance, capacity utilization, safety, and design functionality. Factors are themselves considered to be composed of indicators, v_i, expressed as

$$x_i\left(v_1, v_2, \ldots, v_n\right) = \sum_{j=1}^{n} z_i\left(v_i\right)$$

where $\{v_j\}$ = set of n indicators associated with factor $x_i\left(i = 1, 2, \ldots, m\right)$ and z_j = scaling function for each indicator variable v_j. Examples of indicators are project responsiveness, lead time, learning curve (Badiru, 2010), and work rejects (Badiru, 2014). By combining the above definitions, a composite measure of the value of a process can be modeled as:

$$V = f\left(A_1, A_2, \ldots, A_p\right)$$

$$= f\left\{\left[\sum_{i=1}^{m_1} f_1\left(\sum_{j=1}^{n} z_j\left(v_j\right)\right)\right]_1, \left[\sum_{i=1}^{m_2} f_2\left(\sum_{j=1}^{n} z_j\left(v_j\right)\right)\right]_2, \ldots, \left[\sum_{i=1}^{m_p} f_p\left(\sum_{j=1}^{n} z_j\left(v_j\right)\right)\right]_p\right\}$$

where m and n may assume different values for each attribute. A subjective measure to indicate the utility of the decision maker may be included in the model by using an attribute weighting factor, w_i, to obtain a weighted PV:

$$PV_w = f\left(w_1 A_1, w_2 A_2, \ldots, w_p A_p\right)$$

where

$$\sum_{k=1}^{p} w_k = 1, \qquad \left(0 \le w_k \le 1\right)$$

With this modeling approach, a set of process options can be compared on the basis of a set of attributes and factors. To illustrate the model above, suppose three IT options are to be evaluated based on four attribute elements: *capability, suitability, performance,* and *productivity* (see Table 3.1). For this example, based on the above equations, the value vector is defined as:

$$V = f\left(capability, suitability, performance, productivity\right)$$

Capability: The term "capability" refers to the ability of IT equipment to satisfy multiple requirements. For example, a certain piece of IT equipment may only provide computational service. A different piece of equipment may be capable of generating reports in addition to computational analysis, thus increasing the service variety that can be obtained. In Table 3.1, the levels of increase in service variety from the three competing equipment types are 38%, 40%, and 33%, respectively. *Suitability:* "Suitability" refers to the appropriateness of the IT equipment for current operations. For example, the respective percentages of operating scope for which the three options are suitable for are 12%, 30%, and 53%. *Performance:* "Performance," in this context, refers to the ability of the IT equipment to satisfy schedule and cost requirements. In the example, the three options can, respectively, satisfy requirements on 18%, 28%, and 52% of the typical set of jobs. *Productivity:* "Productivity" can be measured by an assessment of the performance of the proposed IT equipment to meet workload requirements in relation to the existing equipment. For example in Table 3.1, the three options, respectively, show normalized increases of 0.02, -1.0, and -1.1 on a uniform scale of productivity measurement. A plot of the histograms of

---

TABLE 3.1

Comparison of IT Work Value Options

IT Equipment Options	Suitability (k = 1)	Capability (k = 2)	Performance (k = 3)	Productivity (k = 4)
Option A	0.12	0.38	0.18	0.02
Option B	0.30	0.40	0.28	−1.00
Option C	0.53	0.33	0.52	−1.10

the respective "values" of the three IT options were evaluated to find Option C as the best "value" alternative in terms of suitability and performance. Option B shows the best capability measure, but its productivity is too low to justify the needed investment. Option A offers the best productivity, but its suitability measure is low. The analytical process can incorporate a lower control limit into the quantitative assessment such that any option providing value below that point will not be acceptable. Similarly, a minimum value target can be incorporated into the graphical plot such that each option is expected to exceed the target point on the value scale.

The relative weights used in many justification methodologies are based on subjective propositions of decision makers. Some of those subjective weights can be enhanced by the incorporation of utility models. For example, the weights shown in Table 3.1 could be obtained from utility functions. There is a risk of spending too much time maximizing inputs at "point-of-sale" levels with little time defining and refining outputs at the "wholesale" systems level. Without a systems view, we cannot be sure if we are pursuing the right outputs.

SOFT SKILLS FOR LEADERSHIP

Project management continues to grow as an effective means of managing functions in any organization. Project management should be an enterprise-wide systems-based endeavor. Enterprise-wide project management is the application of project management techniques and practices across the full scope of the enterprise. This concept is also referred to as management by project (MBP). MBP is a contemporary concept that employs project management techniques in various functions within an organization. MBP recommends pursuing endeavors as project-oriented activities. It is an effective way to conduct any business activity. It represents a disciplined approach that defines any work assignment as a project. Under MBP, every undertaking is viewed as a project that must be managed just like a traditional project. The characteristics required of each project so defined are:

1. An identified scope and a goal
2. A desired completion time
3. Availability of resources
4. A defined performance measure
5. A measurement scale for review of work

An MBP approach to operations helps in identifying unique entities within functional requirements. This identification helps determine where functions overlap and how they are interrelated, thus paving the way for better planning, scheduling, and control. Enterprise-wide project management facilitates a unified view of organizational goals and provides a way for project teams to use information generated by other departments to carry out their functions.

The use of project management continues to grow rapidly. The need to develop effective management tools increases with the increasing complexity of new technologies and processes. The life cycle of a new product to be introduced into a competitive market is a good example of a complex process that must be managed with integrative project management approaches. The product will encounter management functions as it goes from one stage to the next. Project management will be needed throughout the design and production stages of the product. Project management will be needed in developing marketing, transportation, and delivery strategies for the product. When the product finally gets to the customer, project management will be needed to integrate its use with those of other products within the customer's organization.

The need for a project management approach is established by the fact that a project will always tend to increase in size even if its scope is narrowing. The following four literary laws are applicable to any project environment:

Parkinson's law: Work expands to fill the available time or space.
Peter's principle: People rise to the level of their incompetence.
Murphy's law: Whatever can go wrong will.
Badiru's rule: The grass is always greener where you most need it to be dead.

An integrated systems project management approach can help diminish the adverse impacts of these laws through good project planning, organizing, scheduling, and controlling.

INTEGRATED SYSTEMS IMPLEMENTATION

Project management tools can be classified into three major categories:

1. Qualitative tools. These are the managerial tools that aid in the interpersonal and organizational processes required for project management.
2. Quantitative tools. These are analytical techniques that aid in the computational aspects of project management.
3. Computer tools. These are software and hardware tools that simplify the process of planning, organizing, scheduling, and controlling a project. Software tools can help in both the qualitative and quantitative analyses needed for project management.

Although individual books dealing with management principles, optimization models, and computer tools are available, there are few guidelines for the integration of the three areas for project management purposes. In this book, we integrate these three areas for a comprehensive guide to project management. The book introduces the *Triad Approach* to improve the effectiveness of project management with respect to schedule, cost, and performance constraints within the context of systems modeling. Figure 3.9 illustrates this emphasis from a work systems management perspective. The approach considers not only the management of the work itself but also the management of all the worker-related functions that support the work.

It is one thing to have a quantitative model, but it is a different thing to be able to apply the model to real-world problems in a practical form. The systems approach presented in this book illustrates how to make the transition from model to practice.

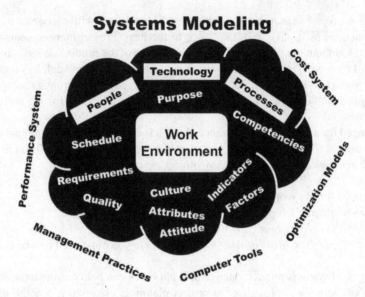

FIGURE 3.9 Project systems modeling in the work environment.

A system approach helps increase the intersection of the three categories of project management tools and, hence, improve overall management effectiveness. Crisis should not be the instigator for the use of project management techniques. Project management approaches should be used upfront to prevent avoidable problems rather than to fight them when they develop. What is worth doing is worth doing well, right from the beginning.

FACTORS OF EFFICIENCY AND EFFECTIVENESS

The premise of this book is that the critical factors for systems success revolve around people and the personal commitment and dedication of each person. No matter how good a technology is and no matter how enhanced a process might be, it is ultimately the people involved that determine the success. This makes it imperative to take care of people's issues first in the overall systems approach to project management. Many organizations recognize this, but only few have been able to actualize the ideals of managing people productively. Execution of operational strategies requires forthrightness, openness, and commitment to get things done. Lip service and arm waving are not sufficient. Tangible programs that cater to the needs of people must be implemented. It is essential to provide incentives, encouragement, and empowerment for people to be self-actuating in determining how best to accomplish their job functions. A summary of critical factors for systems success encompasses the following:

Total system management: hardware, software, and people

- Operational effectiveness
- Operational efficiency
- System suitability

- System resilience
- System affordability
- System supportability
- System life cycle cost
- System performance
- System schedule
- System cost

Systems engineering tools, techniques, and processes are essential for project life cycle management to make goals possible within the context of *SMART* principles, which are represented as follows:

1. Specific: Pursue specific and explicit outputs
2. Measurable: Design of outputs that can be tracked, measured, and assessed
3. Achievable: Make outputs to be achievable and aligned with organizational goals
4. Realistic: Pursue only the goals that are realistic and result-oriented
5. Timed: Make outputs timed to facilitate accountability

Systems engineering provides the technical foundation for executing a project successfully. A systems approach is particularly essential in the early stages of the project in order to avoid having to re-engineer the project at the end of its life cycle. Early systems engineering makes it possible to proactively assess feasibility of meeting user needs, adaptability of new technology, and integration of solutions into regular operations.

SYSTEMS HIERARCHY FOR HUMAN WORK

The traditional concepts of systems analysis are applicable to the project process. The definitions of a project system and its components are presented next.

SYSTEM

A project system consists of interrelated elements organized for the purpose of achieving a common goal. The elements are organized to work synergistically to generate a unified output that is greater than the sum of the individual outputs of the components.

PROGRAM

A program is a very large and prolonged undertaking. Such endeavors often span several years. Programs are usually associated with particular systems. For example, we may have a space exploration program within a national defense system.

PROJECT

A project is a time-phased effort of much smaller scope and duration than a program. Programs are sometimes viewed as consisting of a set of projects. Government projects are often called *programs* because of their broad and comprehensive nature. Industry tends to use the term *project* because of the short-term and focused nature of most industrial efforts.

TASK

A task is a functional element of a project. A project is composed of a sequence of tasks that all contribute to the overall project goal.

ACTIVITY

An activity can be defined as a single element of a project. Activities are generally smaller in scope than tasks. In a detailed analysis of a project, an activity may be viewed as the smallest, practically indivisible work element of the project. For example, we can regard a manufacturing plant as a system. A plant-wide endeavor to improve productivity can be viewed as a program. The installation of a flexible manufacturing system is a project within the productivity improvement program. The process of identifying and selecting equipment vendors is a task, and the actual process of placing an order with a preferred vendor is an activity.

The emergence of systems development has had an extensive effect on project management in recent years. A system can be defined as a collection of interrelated elements brought together to achieve a specified objective. In a management context, the purposes of a system are to develop and manage operational procedures and to facilitate an effective decision-making process. Some of the common characteristics of a system include:

1. Interaction with the environment
2. Objective
3. Self-regulation
4. Self-adjustment

Representative components of a project system are the organizational subsystem, planning subsystem, scheduling subsystem, information management subsystem, control subsystem, and project delivery subsystem. The primary responsibilities of project analysts involve ensuring the proper flow of information throughout the project system. The classical approach to the decision process follows rigid lines of organizational charts. By contrast, the systems approach considers all the interactions necessary among the various elements of an organization in the decision process.

The various elements (or subsystems) of the organization act simultaneously in a separate but interrelated fashion to achieve a common goal. This synergism helps to expedite the decision process and to enhance the effectiveness of decisions. The supporting commitments from other subsystems of the organization serve to counterbalance the weaknesses of a given subsystem. Thus, the overall effectiveness of the system is greater than the sum of the individual results from the subsystems.

SYSTEMS INTEGRATION FOR PROJECT MANAGEMENT

The increasing complexity of organizations and projects makes the systems approach essential in today's management environment. As the number of complex projects increase, there will be an increasing need for project management professionals who can function as systems integrators. Project management techniques can be applied to the various stages of implementing a system as shown in the following guidelines:

1. Systems definition. Define the system and associated problems using keywords that signify the importance of the problem to the overall organization. Locate experts in this area who are willing to contribute to the effort. Prepare and announce the development plan.
2. Personnel assignment. The project group and the respective tasks should be announced, a qualified project manager should be appointed, and a solid line of command should be established and enforced.
3. Project initiation. Arrange an organizational meeting during which a general approach to the problem should be discussed. Prepare a specific development plan and arrange for the installation of needed hardware and tools.
4. System prototype. Develop a prototype system, test it, and learn more about the problem from the test results.
5. Full system development. Expand the prototype to a full system, evaluate the user interface structure, and incorporate user training facilities and documentation.
6. System verification. Get experts and potential users involved, ensure that the system performs as designed, and debug the system as needed.
7. System validation. Ensure that the system yields expected outputs. Validate the system by evaluating performance level, such as percentage of success in so many trials, measuring the level of deviation from expected outputs, and measuring the effectiveness of the system output in solving the problem.
8. System integration. Implement the full system as planned, ensure the system can coexist with systems already in operation, and arrange for technology transfer to other projects.
9. System maintenance. Arrange for continuing maintenance of the system. Update solution procedures as new pieces of information become available. Retain responsibility for system performance or delegate to well-trained and authorized personnel.
10. Documentation. Prepare full documentation of the system, prepare a user's guide, and appoint a user consultant.

Systems integration permits sharing of resources. Physical equipment, concepts, information, and skills may be shared as resources. Systems integration is now a major concern of many organizations. Even some of the organizations that traditionally compete and typically shun cooperative efforts are beginning to appreciate the value of integrating their operations. For these reasons, systems integration has emerged as a major interest in business. Systems integration may involve the physical integration of technical components, objective integration of operations, conceptual integration of management processes, or a combination of any of these.

Systems integration involves the linking of components to form subsystems and the linking of subsystems to form composite systems within a single department and/

or across departments. It facilitates the coordination of technical and managerial efforts to enhance organizational functions, reduce cost, save energy, improve productivity, and increase the utilization of resources. Systems integration emphasizes the identification and coordination of the interface requirements among the components in an integrated system. The components and subsystems operate synergistically to optimize the performance of the total system. Systems integration ensures that all performance goals are satisfied with a minimum expenditure of time and resources. Integration can be achieved in several forms including the following:

1. Dual-use integration: This involves the use of a single component by separate subsystems to reduce both the initial cost and the operating cost during the project life cycle.
2. Dynamic resource integration: This involves integrating the resource flows of two normally separate subsystems so that the resource flow from one to or through the other minimizes the total resource requirements in a project.
3. Restructuring of functions: This involves the restructuring of functions and reintegration of subsystems to optimize costs when a new subsystem is introduced into the project environment.

Systems integration is particularly important when introducing new work into an existing system. It involves coordinating new operations to coexist with existing operations. It may require the adjustment of functions to permit the sharing of resources, development of new policies to accommodate product integration, or realignment of managerial responsibilities. It can affect both hardware and software components of an organization. Presented below are guidelines and important questions relevant for work systems integration.

- What are the unique characteristics of each component in the integrated system?
- How do the characteristics complement one another?
- What physical interfaces exist among the components?
- What data/information interfaces exist among the components?
- What ideological differences exist among the components?
- What are the data flow requirements for the components?
- Are there similar integrated systems operating elsewhere?
- What are the reporting requirements in the integrated system?
- Are there any hierarchical restrictions on the operations of the components of the integrated system?
- What internal and external factors are expected to influence the integrated system?
- How can the performance of the integrated system be measured?
- What benefit/cost documentations are required for the integrated system?
- What is the cost of designing and implementing the integrated system?
- What are the relative priorities assigned to each component of the integrated system?
- What are the strengths of the integrated system?

- What are the weaknesses of the integrated system?
- What resources are needed to keep the integrated system operating satisfactorily?
- Which section of the organization will have primary responsibility for the operation of the integrated system?
- What are the quality specifications and requirements for the integrated systems?

The effectiveness of using the DEJI Systems Model rests in the model's ability to harmonize new work with existing work or work processes. The model's efficacy can be seen in the following selected examples, which highlight how things are expected fit together.

- Administrative integration
- Technical integration
- Behavioral integration
- Operational integration
- Philosophical integration
- Religious integration
- Economic integration
- Financial integration
- Security integration
- Safety integration
- Social integration

This approach guides leaders to explicit assess where and how work elements and processes harmonize (i.e., integrate).

MANAGING WORKERS' HIERARCHY OF NEEDS

Maslow's Hierarchy of Needs is very much applicable in any work environment.

Abraham Maslow (Maslow, 1943), a psychologist, is best known for his theory of the "Hierarchy of Needs," which categorized the basic human needs that must be met before an individual can seek social or spiritual fulfillment. Feeling that psychology didn't take into account human creativity or potential, Maslow defined the concept of "self-actualization" as a process in which humans continually strive to reach the best level of personal accomplishment. Personal choice played a prominent part in Maslow's theories, which contend that human progress in life is up to individuals, if we have the fortitude to move forward into the uncharted areas. An individual will either advance forward into the areas of growth or will step backward into the lowest level of needs (i.e., safety level). According to Maslow's theory (Maslow, 1943), the five different orders of human needs are:

1. Basic physiological needs: It includes food, water, shelter, and the like. In modern society, the basic drives of human existence cause individuals to become involved in organizational life. People become participants in the

organization that employs them. Thus, at the simplest level of human needs, people are motivated to join organizations, remain in them, and contribute to their objectives.

2. Security and safety: Security means many things to different people in different circumstances. For some, it means earning a higher income to assure freedom from what might happen in case of sickness or during old age. Thus many people are motivated to work harder to seek success which is measured in terms of income. It can also be interpreted as job security. To some people such as civil servants and teachers, the assurance of life tenure and a guaranteed pension may be strong motivators in their participation in employing organizations.

3. Social affiliation: An employee with a reasonable well-paying and secure job will begin to feel that belonging and approval are important motivators in his/her organizational behavior.

4. Esteem: The need to be recognized, to be respected, and to have prestige (self-image and the view that one holds of oneself). There is a dynamic interplay between one's own sense of satisfaction and self-confidence on one hand, and feedback from others in such diverse forms as being asked for advice on the other hand.

5. Self-actualization: The desire to become more and more what one is, to become everything that one is capable of becoming. The self-actualized person is strongly inner-directed, seeks self-growth, and is highly motivated by loyalty to cherished values, ethics, and beliefs. Not everyone reaches the self-actualized state. It is estimated that these higher-level needs are met about 10% of the time.

In any organization, the prevailing hierarchy of needs of the worker must be evaluated in the context of the organization's own hierarchy of needs. In this regard, Badiru (2008) developed an adaptation of the conventional triangle of the hierarchy into a multi-dimensional pyramid of needs as illustrated in Figure 3.10.

LEADING ECONOMIC DEVELOPMENT

Not only is work essential for personal and organizational advancement, it is also essential, from a synergistic systems perspective, for national social and economic development. The gross domestic product (GDP) is the eventual coalescing of work done at various levels of the nation. GDP is the monetary value of all the finished goods and services produced within a country in a defined period of time. Though GDP is usually calculated on an annual basis, it can be calculated on a quarterly basis as well in order to increase the granularity of management policies and practices to increase the national output. GDP includes all private and public consumption, government spending, investments, and exports (minus imports) that occur within a national boundary. In other words, GDP is a broad and composite measurement of a nation's overall economic activity. It can be calculated as follows:

$$GDP = C + G + I + NX$$

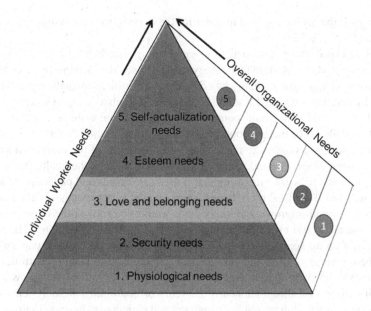

FIGURE 3.10 Multidimensional pyramid of needs of workers and the organization.

where

C = all private consumption, or consumer spending

G = the sum of government spending

I = sum of all the country's investment, including corporate capital expenditures

NX = the nation's total net exports, calculated as total exports minus total imports

GDP is commonly used as an indicator of the economic health of a country, as well as a gauge of a country's standard of living. If a country's standard of living is high, the workers in the country do well for themselves. Since the mode of measuring GDP is uniform from country to country, GDP can be used to compare the productivity of various countries with a high degree of accuracy. Adjusting for inflation from year to year allows for an objective comparison of current GDP measurement trends. Thus, a country's GDP from any period can be measured as a percentage relative to previous years or quarters. Consequently, GDP trends can be used in measuring a nation's economic growth or decline, as well as in determining if an economy is in recession, which has a direct impact on workers and the work environment.

LEADING A TEAM OF TEAMS

The DEJI Systems Model is aptly applicable for leading a Team of Teams (TOT), in alignment with the principles of System of Systems (SOS). Each team is a subset of a larger team in the way that each system is a subsystem of a larger system. Using the Triple C principle and DEJI Systems Model, communication and integration must permeate all the layers involved in order for the overall team (or system) to reach its intended level of effectiveness.

SOS emerged as an approach to leverage the power behind disparate systems operating as a whole. Modern technology made this science possible by offering new ways to connect, interoperate, model complexity, examine, and analyze possible outcomes by running simulated test and evaluation. SOS also emerged when the problems of our world became quite complex and the individual systems we created were not sufficient to solve them. People began postulating on the benefits of combining previously separate systems to render outcomes that none of the individual systems could produce alone. A good example of a modern SOS is the healthcare system. Healthcare leaders began imagining that if medical records and patient portals were connected, and if medical records were made available between hospitals, that care for patients would improve, and likely costs would go down as the interoperability of the systems would eliminate manual operation of certain functions. This would notionally free up healthcare workers from the drudgery of medical paperwork to have them give more time and care to their patients and to keep up with the latest developments in their chosen fields of medicine. In theory, this outcome was plausible. It can be said that after nearly 20 years of combining these systems, many benefits and outcomes have not been realized and patient care has not improved. The point is that a SOS can be imagined and a case for it can be made but realizing such a "beast" is entirely another matter.

Where did the problems emerge? Was the combining of these systems a kluge rather than a well-thought-out design? Were there competing interests among all the stakeholders that were not considered? Was it postulated that technology would be easy to implement on a grand scale and "solve all ills," only to have it fall short in many areas? The answers to these questions will likely be sought after for years (decades?) to come. In the meantime, it is important for leaders in any field of study to tread cautiously when applying the SOS approach, methodology, thinking (wishing!) to individual units of operation in the hopes of "better outcomes." Anticipating the emergent behaviors of the whole is a daunting task, as the individual units were designed never intending to do anything more than what they were designed to do (separately and independently). Now they are being asked to interoperate, collaborate, coordinate, subordinate (think: "I can't do that, Dave" 2001 – A Space Odyssey).

And then we have the people. People are the cornerstone of industry and commerce, making all outcomes possible through their work efforts. The big piece here (post COVID-19 pandemic) is the geographic dispersion of people and the "new normal"; people working from home and away from the workplace. The traditional workplace is "done for." This eliminates the once concrete boundaries that defined people's workplaces and hence the rules, processes, policies, and operations of their jobs. Incidentally, vanishing boundaries and geographic dispersion, it turns out, are two criteria for using an SOS approach (according to INCOSE) to solve problems. If that is so, can an SOS approach be applied to people as they continue to operate in teams that are now made much more complex (another criteria for SOS), reliant on technology (another criteria for SOS), and highly dependent on information exchange across vast distances and time zones? Yes.

As people migrate to teams with a whole new set of boundary conditions, their interoperability will undoubtedly produce emergent outcomes. To cope with this, the entire combination will evolve, adopting new policies and technology to continue the workflow. All of this mimics the intricacies of SOS. The larger the company, the more teams form, and the imposition of hierarchy still exists. Now, we have a TOT performing the operations of what was once disparate units of people, co-located and able to operate due to their sharing space and having their work environments sanitized with cubicles and desk phones and computers (and a token, local IT guy). "The workplace no longer operates in isolation" (Katina and Magpili, 2022). *If TOT is the new normal for the post-pandemic work world, then who is leading such teams?*

Leadership as a practice, a science, an approach, and a theory has been studied heavily in the last 150 years. Much has been published on the subject (Kirpes, 2022; Schall, 2022) and many theorists constructed their frameworks to explain the dynamics of leading, following, motivating, inspiring, planning, and organizing to have the people being led produce and be efficient and happy. Those teams were primarily linear and finite, distinct entities producing their intended outcomes independent of one another. Now, leaders are being expected to lead TOT. But the following questions apply:

- How will they cope? How will they model the complexity of the interactions?
- How can leaders possibly anticipate the emergent behavior, notionally to leverage it for better outcomes?
- How can they manage the vanishing boundaries?
- How can they see to it the latest technology is incorporated into the teams for optimal connectivity and interoperability?
- How can they manage the vast information flows as a result of Globalization?
- How can they handle competing priorities and keep teams on course for the goals of the organization?
- How can they close the "leadership trust gap" within and between teams of teams in a complex organizational structure?

TEAM OF TEAMS APPROACH FOR LEADERSHIP

In this section, we propose a novel approach for leading TOT. It is based on an integrative application of DEJI Systems Model (Badiru, 2023) and the Triple C Principle (Badiru, 2008) in which communication, cooperation, coordination, and integration play crucial anchoring roles of the leader and the led. Further, the approach harnesses the power of System of Systems Engineering (SOSE) and SOS tools, techniques, models, proven best practices, lessons learned. It also anticipates emergent outcomes and see them as enablers, not obstacles, just as SOS does. The novel methodology, which is subject to additional scholarly research, will stimulate and guide leaders to up their game and take on the challenge of managing a TOT. This is particularly constructive for technical professionals (e.g., engineers) rising to the role of leaders (Kaminski, 2023).

CONCLUSION

In summary for this chapter, we reaffirm that soft skills can impact Total Worker Health® (Schill, 2016; Schill and Chosewood, 2013). A healthy workforce is a more productive workforce. It is through the application of soft skills that organizations can guide employees toward benchmarks and expectations for total work health, which then has an impact on organizational efficiency, effectiveness, and productivity. Work, wellness, and wealth can go hand-in-hand. Workers' health directly affects GDP, based on a systems view of work, cascading from one person's level all the way (collectively) to the national level. Good health is related to good work performance. Health is an individual attribute that compliments each worker's hierarchy of needs. Without good health, even the best worker cannot perform. Without good health, even the best athlete cannot succeed. Without good health, even the most proficient expert cannot manifest his or her expertise. Without total health, even the most dedicated and experienced employee cannot contribute to the accomplishment of the organization's mission. Good health is a key part of the system's view of work as advocated by this chapter.

REFERENCES

Agustiady, Tina and Badiru, A. B. (2012) Sustainability: Utilizing Lean Six Sigma Techniques. Taylor & Francis CRC Press, Boca Raton, FL.

Badiru, Adedeji B. (2008) Triple C Model of Project Management. Taylor & Francis CRC Press, Boca Raton, FL.

Badiru, Adedeji B. (2010) Half-Life of Learning Curves for Information Technology Project Management. *International Journal of IT Project Management, 1*(3), 28–45.

Badiru, Adedeji B. (2012) Project Management: Systems, Principles, and Applications. Taylor & Francis CRC Press, Boca Raton, FL.

Badiru, Adedeji B. (2014) Quality Insights: The DEJI Model for Quality Design, Evaluation, Justification, and Integration. *International Journal of Quality Engineering and Technology, 4*(4), 369–378.

Badiru, Adedeji B. (2019) Systems Engineering Models: Theory, Methods, and Applications. Taylor & Francis/CRC Press, Boca Raton, FL.

Badiru, Adedeji B. (2023a), "Soft Skills in Industrial Engineering and Management," Chapter 71 in Maynard's Industrial & Systems Engineering Handbook, Sixth Edition, Bopaya Bidanda, editor. McGraw-Hill, New York, NY, pp. 1383–1400.

Badiru, Adedeji B. (2023b) Systems Engineering Using DEJI Systems Model: Design, Evaluation, Justification, and Integration with Case Studies and Applications. Taylor & Francis CRC Press, Boca Raton, FL.

Badiru, Ibrahim Ade (2016) "Comments about Work Management," Interview of an Auto Industry Senior Engineer about Corporate Views of Work Design, Beavercreek, Ohio, October 29, 2016.

Kaminski, Michael J. (2023) "The Industrial Engineer as a Manager," Chapter 77 in **Maynard's Industrial & Systems Engineering Handbook**, Bopaya Bidanda, editor. McGraw-Hill, New York, NY, pp. 1495–1512.

Katina, Polinpapilinho F., and Magpili, Nina. *System-of-systems to the rescue? Solving unsolvable problems by Polipapilinho Katina, Nina Magpili*. The Market for Ideas. Retrieved December 11, 2022, from https://www.themarketforideas.com/system-of-systems-to-the-rescue-solving-unsolvable-problems-a187/

Kirpes, Cárl (2022) "Thoughts on Servant Leadership," **ISE Magazine**, December 2022, p. 22.

Maslow, Abraham H. (1943) A Theory of Human Motivation. *Psychological Review, 50*, 370–396. https://doi.org/10.1037/h0054346

Schall, Susan O. (2022) "The Essential Leader for Changing Workplaces: Relationships Must Lay the Foundation for a Culture of Improvement," **ISE Magazine**, December 2022, pp. 32–37.

Schill, Anita L. (2016), "Advancing Well-being Through Total Worker Health," Keynote Address, 17th Annual 2016 Pilot Research Project (PRP) Symposium, University of Cincinnati, Cincinnati, OH, October 13, 2016.

Schill, Anita L. and Chosewood, L. C. (2013) The NIOSH Total Worker Health Program: an Overview. *Journal of Occupational and Environmental Medicine, 55*(12 suppl), S8–S11.

Schulte, Paul A., Guerin, Rebecca J., Schill, Anita L., Bhattacharya, Anasua, Cunningham, Thomas R., Pandalai, Sudha P., Eggerth, Donald and Stephenson, Carol M. (2015) Considerations for Incorporating 'Well-Being' in Public Policy for Workers and Workplaces. *American Journal of Public Health, 105*(8), e31–e44.

Troxler, Joel W. and Blank, L. (1989) A Comprehensive Methodology for Manufacturing System Evaluation and Comparison. *Journal of Manufacturing Systems, 8*(3), 176–183.

4 Triple-C Model for Leadership Effectiveness

INTRODUCTION

> Great discoveries and improvements invariably involve the cooperation of many minds.
>
> *Alexander Graham Bell*

This chapter addresses project planning requirements from a systems perspective as a precursor to securing cooperation. Without a plan, there is no basis for cooperation. Cooperation across many minds involving several subsystems of an enterprise is essential for organizational success. The required components of a plan are discussed in this chapter.

Ultimately, cooperation among people in the workforce is what gets the job done. This chapter presents the Triple-C model (Badiru, 2008) for communication, cooperation, and coordination. We must communicate before we can secure cooperation so that we can coordinate individual efforts. As Alexander Bell says in the opening quote of this chapter, improvements (and successes) require cooperation of minds.

On the matter of risk management in a work environment, we should recognize that risk is inherent in every project, whether it is big or small, whether it is avoidable or not. On this note, Park's law quoted below is appropriate. Park's Law of the Conservation of Misery:

> The sum of the misery (or difficulty or risk) cannot be driven lower than a constant. It can be shifted between disciplines, but not taken below the minimum.
>
> *Erik Park*

Based on its leading edge of communication, the Triple-C model can facilitate risk mitigation.

ORIGIN OF THE TRIPLE-C MODEL

The origin of the Triple-C model demonstrates that it is possible to lead from behind if there is a solid plan that everyone can follow. The idea for the Triple-C model originated from a complex facility redesign project (Badiru et al., 1993; Badiru, 2019) conducted for Tinker Air Force Base (TAFB) in Oklahoma City by the School of Industrial Engineering, University of Oklahoma from 1985 through 1989. The project was a part of a reconstruction project following a disastrous fire that occurred in the base's repair/production facility in November 1984. The urgency, complexity, scope ambiguity, confusion, and disjointed directions

DOI: 10.1201/9781003311348-4

that existed in the early days of the reconstruction effort led to the need to develop a structured approach to communication, cooperation, and coordination of the various work elements. In spite of the high pressure timing of the project, the author called a Time-Out-Of-time (TOOT) so that a process could be developed for project communication leading to personnel cooperation, and eventually facilitating task coordination. The investment of TOOT time resulted in a remarkable resurgence of cooperation where none existed at the beginning of the project. Encouraged by the intrinsic occurrence of cooperation, the process was further enhanced and formalized as the Triple-C approach to the project's success. The approach was credited with the overall success of the project. The qualitative approach of Triple-C complemented the technical approaches used on the project to facilitate harmonious execution of tasks. Many projects fail when the stakeholders get too wrapped up into the technical requirements at the expense of qualitative requirements. Other elements of "C," such as Collaboration, Commitment, and Correlation, are embedded in the Triple-C structure. Of course, the constraints of time, cost, and performance must be overcome all along the way.

Organizations thrive by investing in three primary resources: *People* who do the work, the *Tools* that the people use to do the work, and the *Process* that governs the work that the people do. Of the three, investing in people is the easiest thing an organization can do and we should do it whenever we have an opportunity. The Triple-C approach incorporates the essential elements of human interactions, with respect to communication, cooperation, and coordination.

TRIPLE-C MODEL

The Triple-C model was first used in 1985 and subsequently introduced in print in 1987 (Badiru, 1987). The project scenario that led to the development of the Triple-C Model was later documented in Badiru et al. (1993). The model is an effective project planning and control tool. The model states that project management can be enhanced by implementing it within the integrated functions summarized below:

- Communication
- Cooperation
- Coordination

The model facilitates a systematic approach to project planning, organizing, scheduling, and controlling. The Triple-C model is distinguished from the 3C approach commonly used in military operations. The military approach emphasizes personnel management in the hierarchy of command, control, and communication. This places communication as the last function. The Triple-C, by contrast, suggests communication as the first and foremost function. The Triple-C model can be implemented for project planning, scheduling, and controlling purposes.

Figure 4.1 shows the application of Triple-C for project planning, scheduling, and controlling within the confines of the Triple Constraints of cost, schedule, and performance. Each of these three primary functions of project management requires effective communication, sustainable cooperation, and adaptive coordination.

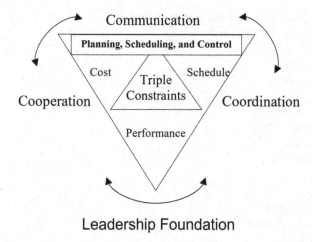

FIGURE 4.1 Triple-C for planning, scheduling, and control.

The basic questions that revolve around an effective implementation of Triple-C are:

- What?
- Who?
- Why?
- How?
- Where?
- When?

The questions highlight what must be done, where, when, why, and how. It can also help to identify the resources (personnel, equipment, facilities, etc.) required for each effort. It points out important questions such as:

- Does each project participant know what the objective is?
- Does each participant know his or her role in achieving the objective?
- What obstacles may prevent a participant from playing his or her role effectively?

Triple-C can mitigate disparity between idea and practice because it explicitly solicits information about the critical aspects of a project.

Types of Communication
- Verbal
- Written
- Body language
- Visual tools (e.g., graphical tools)
- Sensual (Use of all five senses: sight, smell, touch, taste, hearing:- olfactory, tactile, auditory)
- Simplex (unidirectional)

- Half-duplex (bi-directional with time lag)
- Full-duplex (real-time dialogue)
- One-on-one
- One-to-many
- Many-to-one

Types of Cooperation
- Proximity
- Functional
- Professional
- Social
- Romantic
- Power influence
- Authority influence
- Hierarchical
- Lateral
- Cooperation by intimidation
- Cooperation by enticement

Types of Coordination
- Teaming
- Delegation
- Supervision
- Partnership
- Token-passing
- Baton hand-off

TYPICAL TRIPLE-C QUESTIONS

Questioning is the best approach to getting information for effective project management. Everything should be questioned. By upfront questions, we can preempt and avert project problems later on. Typical questions to ask under Triple-C approach are:

- What is the purpose of the project?
- Who is in charge of the project?
- Why is the project needed?
- Where is the project located?
- When will the project be carried out?
- How will the project contribute to increased opportunities for the organization?
- What is the project designed to achieve?
- How will the project affect different groups of people within the organization?
- What will be the project approach or methodology?
- What other groups or organizations will be involved (if any)?
- What will happen at the end of the project?
- How will the project be tracked, monitored, evaluated, and reported?

- What resources are required?
- What are the associated costs of the required resources?
- How do the project objectives fit the goal of the organization?
- What respective contribution is expected from each participant?
- What level of cooperation is expected from each group?
- Where is the coordinating point for the project?

The key to getting everyone on board with a project is to ensure that task objectives are clear and comply with the principle of *SMART* as outlined below:

Specific: Task objective must be specific.
Measurable: Task objective must be measurable.
Aligned: Task objective must be achievable and aligned with overall project goal.
Realistic: Task objective must be realistic and relevant to the organization.
Timed: Task objective must have a time basis.

If a task has the above intrinsic characteristics, then the function of communicating the task will more likely lead to personnel cooperation.

COMMUNICATION

What we have here is a failure to communicate.

Common regret in project execution

Communication makes working together possible. The communication function of project management involves making all those concerned become aware of project requirements and progress. Those who will be affected by the project directly or indirectly, as direct participants or as beneficiaries, should be informed as appropriate regarding the following:

- Scope of the project
- Personnel contribution required
- Expected cost and merits of the project
- Project organization and implementation plan
- Potential adverse effects if the project should fail
- Alternatives, if any, for achieving the project goal
- Potential direct and indirect benefits of the project

The communication channel must be kept open throughout the project life cycle. In addition to internal communication, appropriate external sources should also be consulted. The project manager must:

- Exude commitment to the project
- Utilize the communication responsibility matrix
- Facilitate multi-channel communication interfaces

- Identify internal and external communication needs
- Resolve organizational and communication hierarchies
- Encourage both formal and informal communication links

When clear communication is maintained between management and employees and among peers, many project problems can be averted. Project communication may be carried out in one or more of the following formats:

- One-to-many
- One-to-one
- Many-to-one
- Written and formal
- Written and informal
- Oral and formal
- Oral and informal
- Nonverbal gestures

Good communication is affected when what is implied is perceived as intended. Effective communications are vital to the success of any project. Despite the awareness that proper communications form the blueprint for project success, many organizations still fail in their communications functions. The study of communication is complex. Factors that influence the effectiveness of communication within a project organization structure include the following:

1. **Personal perception.** Each person perceives events on the basis of personal psychological, social, cultural, and experimental background. As a result, no two people can interpret a given event the same way. The nature of events is not always the critical aspect of a problem situation. Rather, the problem is often the different perceptions of the different people involved.
2. **Psychological profile.** The psychological makeup of each person determines personal reactions to events or words. Thus, individual needs and level of thinking will dictate how a message is interpreted.
3. **Social environment.** Communication problems sometimes arise because people have been conditioned by their prevailing social environment to interpret certain things in unique ways. Vocabulary, idioms, organizational status, social stereotypes, and economic situation are among the social factors that can thwart effective communication.
4. **Cultural background.** Cultural differences are among the most pervasive barriers to project communications, especially in today's multinational organizations. Language and cultural idiosyncrasies often determine how communication is approached and interpreted.
5. **Semantic and syntactic factors.** Semantic and syntactic barriers to communications usually occur in written documents. Semantic factors are those that relate to the intrinsic knowledge of the subject of the communication. Syntactic factors are those that relate to the form in which the communication is presented. The problems created by these factors become acute

in situations where response, feedback, or reaction to the communication cannot be observed.

6. **Organizational structure.** Frequently, the organization structure in which a project is conducted has a direct influence on the flow of information and, consequently, on the effectiveness of communication. Organization hierarchy may determine how different personnel levels perceive a given communication.

7. **Communication media.** The method of transmitting a message may also affect the value ascribed to the message and consequently, how it is interpreted or used. The common barriers to project communications are:

- Inattentiveness
- Lack of organization
- Outstanding grudges
- Preconceived notions
- Ambiguous presentation
- Emotions and sentiments
- Lack of communication feedback
- Sloppy and unprofessional presentation
- Lack of confidence in the communicator
- Lack of confidence by the communicator
- Low credibility of communicator
- Unnecessary technical jargon
- Too many people involved
- Untimely communication
- Arrogance or imposition
- Lack of focus

Some suggestions on improving the effectiveness of communication are presented next. The recommendations may be implemented as appropriate for any of the forms of communications listed earlier. The recommendations are for both the communicator and the audience.

1. Never assume that the integrity of the information sent will be preserved as the information passes through several communication channels. Information is generally filtered, condensed, or expanded by the receivers before relaying it to the next destination. When preparing a communication that needs to pass through several organization structures, one safeguard is to compose the original information in a concise form to minimize the need for re-composition of the project structure.

2. Give the audience a central role in the discussion. A leading role can help make a person feel a part of the project effort and responsible for the projects' success. He or she can then have a more constructive view of project communication.

3. Do homework and think through the intended accomplishment of the communication. This helps eliminate trivial and inconsequential communication efforts.

4. Carefully plan the organization of the ideas embodied in the communication. Use indexing or points of reference whenever possible. Grouping ideas into related chunks of information can be particularly effective. Present the short messages first. Short messages help create focus, maintain interest, and prepare the mind for the longer messages to follow.

5. Highlight why the communication is of interest and how it is intended to be used. Full attention should be given to the content of the message with regard to the prevailing project situation.

6. Elicit the support of those around you by integrating their ideas into the communication. The more people feel they have contributed to the issue, the more expeditious they are in soliciting the cooperation of others. The effect of the multiplicative rule can quickly garner support for the communication purpose.

7. Be responsive to the feelings of others. It takes two to communicate. Anticipate and appreciate the reactions of members of the audience. Recognize their operational circumstances and present your message in a form they can relate to.

8. Accept constructive criticism. Nobody is infallible. Use criticism as a springboard to higher communication performance.

9. Exhibit interest in the issue in order to arouse the interest of your audience. Avoid delivering your messages as a matter of a routine organizational requirement.

10. Obtain and furnish feedback promptly. Clarify vague points with examples.

11. Communicate at the appropriate time, at the right place, to the right people.

12. Reinforce words with positive action. Never promise what cannot be delivered. Value your credibility.

13. Maintain eye contact in oral communication and read the facial expressions of your audience to obtain real-time feedback.

14. Concentrate on listening as much as speaking. Evaluate both the implicit and explicit meanings of statements.

15. Document communication transactions for future references.

16. Avoid asking questions that can be answered yes or no. Use relevant questions to focus the attention of the audience. Use questions that make people reflect upon their words, such as, "How do you think this will work?" compared to "Do you think this will work?"

17. Avoid patronizing the audience. Respect their judgment and knowledge.

18. Speak and write in a controlled tempo. Avoid emotionally charged voice inflections.

19. Create an atmosphere for formal and informal exchange of ideas.

20. Summarize the objectives of the communication and how they will be achieved.

Figure 4.2 shows an example of a design of a communication responsibility matrix. A communication responsibility matrix shows the linking of sources of communication and targets of communication. Cells within the matrix indicate the subject of the desired communication. There should be at least one filled cell in each row and each column of the matrix. This assures that each individual of a department has at least

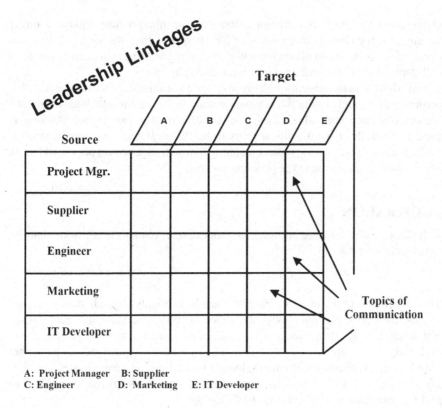

A: Project Manager   B: Supplier
C: Engineer   D: Marketing   E: IT Developer

FIGURE 4.2   Triple-C communication matrix.

one communication source or target associated with him or her. With a communication responsibility matrix, a clear understanding of what needs to be communicated to whom can be developed. Communication in a project environment can take any of the several forms. The specific needs of a project may dictate the most appropriate mode. Three popular computer communication modes are discussed next in the context of communicating data and information for project management.

Simplex communication. This is a unidirectional communication arrangement in which one project entity initiates communication to another entity or individual within the project environment. The entity addressed in the communication does not have mechanism or capability for responding to the communication. An extreme example of this is a one-way, top-down communication from top management to the project personnel. In this case, the personnel have no communication access or input to top management. A budget-related example is a case where top management allocates budget to a project without requesting and reviewing the actual needs of the project. Simplex communication is common in authoritarian organizations.

Half-duplex communication. This is a bi-directional communication arrangement whereby one project entity can communicate with another entity and receive a response within a certain time lag. Both entities can communicate with each other but not at the same time. An example of half-duplex communication is a project organization that permits communication with top management without a direct meeting. Each communicator must wait for a response from the target of the communication. Request and allocation without a budget meeting is another example of half-duplex data communication in project management.

Full-duplex communication. This involves a communication arrangement that permits a dialogue between the communicating entities. Both individuals and entities can communicate with each other at the same time or face-to-face. As long as there is no clash of words, this appears to be the most receptive communication mode. It allows participative project planning in which each project personnel has an opportunity to contribute to the planning process.

COOPERATION

If you want to be incrementally better, be competitive. If you want to be exponentially better, be cooperative.

Ash Lamb on Twitter

The cooperation of the project personnel must be explicitly elicited. Merely voicing consent for a project is not enough assurance of full cooperation. The participants and beneficiaries of the project must be convinced of the merits of the project. Some of the factors that influence cooperation in a project environment include personnel requirements, resource requirements, budget limitations, past experiences, conflicting priorities, and lack of uniform organizational support. A structured approach to seeking cooperation should clarify the following:

- Cooperative efforts required
- Precedents for future projects
- Implication of lack of cooperation
- Criticality of cooperation to project success
- Organizational impact of cooperation
- Time frame involved in the project
- Rewards of good cooperation

Cooperation is a basic virtue of human interaction. More projects fail due to a lack of cooperation and commitment than any other project factors. To secure and retain the cooperation of project participants, you must elicit a positive first reaction to the project. The most positive aspects of a project should be the first items of project communication. For project management, there are different types of cooperation that should be understood.

Functional cooperation. This is cooperation induced by the nature of the functional relationship between two groups. The two groups may be required to perform related functions that can only be accomplished through mutual cooperation.

Social cooperation. This is the type of cooperation effected by the social relationship between two groups. The prevailing social relationship motivates cooperation that may be useful in getting project work done.

Legal cooperation. Legal cooperation is the type of cooperation that is imposed through some authoritative requirement. In this case, the participants may have no choice other than to cooperate.

Administrative cooperation. This is cooperation brought on by administrative requirements that make it imperative that two groups work together on a common goal.

Associative cooperation. This type of cooperation may also be referred to as collegiality. The level of cooperation is determined by the association that exists between two groups.

Proximity cooperation. Cooperation due to the fact that two groups are geographically close is referred to as proximity cooperation. Being close makes it imperative that the two groups work together.

Dependency cooperation. This is cooperation caused by the fact that one group depends on another group for some important aspect. Such dependency is usually of a mutual two-way nature. One group depends on the other for one thing while the latter group depends on the former for some other thing.

Imposed cooperation. In this type of cooperation, external agents must be employed to induce cooperation between two groups. This is applicable for cases where the two groups have no natural reason to cooperate. This is where the approaches presented earlier for seeking cooperation can become very useful.

Lateral cooperation. Lateral cooperation involves cooperation with peers and immediate associates. Lateral cooperation is often easy to achieve because existing lateral relationships create an environment that is conducive for project cooperation.

Vertical cooperation. Vertical or hierarchical cooperation refers to cooperation that is implied by the hierarchical structure of the project. For example, subordinates are expected to cooperate with their vertical superiors.

Whichever type of cooperation is available in a project environment, the cooperative forces should be channeled toward achieving project goals. Documentation of the prevailing level of cooperation is useful for winning further support for a project. Clarification of project priorities will facilitate personnel cooperation. Relative priorities of multiple projects should be specified to declare a unified priority scheme for all groups within the organization.

Some guidelines for securing cooperation for most projects are:

- Establish achievable goals for the project.
- Clearly outline the individual commitments required.
- Integrate project priorities with existing priorities.
- Eliminate the fear of job loss due to industrialization.
- Anticipate and eliminate potential sources of conflict.
- Use an open-door policy to address project grievances.
- Remove skepticism by documenting the merits of the project.

Commitment. Cooperation must be supported with commitment. To cooperate is to support the ideas of a project. To commit is to willingly and actively participate in project efforts again and again through the thick and thin of the project. Provision of resources is one way that management can express commitment to a project.

$$(\text{Triple-C} + \text{Commitment} = \text{Project Success})$$

COORDINATION

After the communication and cooperation functions have successfully been initiated, the efforts of the project personnel must be coordinated. Coordination facilitates harmonious organization of project efforts. The construction of a responsibility chart can be very helpful at this stage. A responsibility chart is a matrix consisting of columns of individual or functional departments and rows of required actions. Cells within the matrix are filled with relationship codes that indicate who is responsible for what. The matrix helps avoid neglecting crucial communication requirements and obligations. It can help resolve questions such as:

- Who is to do what?
- How long will it take?
- Who is to inform whom of what?
- Whose approval is needed for what?
- Who is responsible for which results?
- What personnel interfaces are required?
- What support is needed from whom and when?

CONFLICT RESOLUTION USING TRIPLE-C APPROACH

Conflicts can and do develop in any work environment. Conflicts, whether intended or inadvertent, prevent an organization from getting the most out of the work force. When implemented as an integrated process, the Triple-C model can help avoid conflicts in a project. When conflicts do develop, it can help in resolving the conflicts. The key to conflict resolution is open and direct communication, mutual cooperation, and sustainable coordination. Several sources of conflicts can exist in a projects. Some of these are discussed below.

Schedule conflict. Conflicts can develop because of improper timing or sequencing of project tasks. This is particularly common in large multiple projects. Procrastination can lead to having too much to do at once, thereby creating a clash of project functions and discord among project team members. Inaccurate estimates of time requirements may lead to infeasible activity schedules. Project coordination can help avoid schedule conflicts.

Cost conflict. Project cost may not be generally acceptable to the clients of a project. This will lead to project conflict. Even if the initial cost of the project is acceptable, a lack of cost control during implementation can lead to conflicts. Poor budget allocation approaches and the lack of a financial feasibility study will cause cost conflicts later on in a project. Communication and coordination can help prevent most of the adverse effects of cost conflicts.

Performance conflict. If clear performance requirements are not established, performance conflicts will develop. Lack of clearly defined performance standards can lead each person to evaluate his or her own performance based on personal value judgments. In order to uniformly evaluate quality of work and monitor project progress, performance standards should be established by using the Triple-C approach.

Management conflict. There must be a two-way alliance between management and the project team. The views of management should be understood by the team. The views of the team should be appreciated by the management. If this does not happen, management conflicts will develop. A lack of a two-way interaction can lead to strikes and industrial actions, which can be detrimental to project objectives. The Triple-C approach can help create a conducive dialogue environment between management and the project team.

Technical conflict. If the technical basis of a project is not sound, technical conflict will develop. New industrial projects are particularly prone to technical conflicts because of their significant dependence on technology. Lack of a comprehensive technical feasibility study will lead to technical conflicts. Performance requirements and systems specifications can be integrated through the Triple-C approach to avoid technical conflicts.

Priority conflict. Priority conflicts can develop if project objectives are not defined properly and applied uniformly across a project. Lack of a direct project definition can lead each project member to define his or her own goals which may be in conflict with the intended goal of a project. Lack of consistency of the project mission is another potential source of priority conflicts. Over-assignment of responsibilities with no guidelines for relative significance levels can also lead to priority conflicts. Communication can help defuse priority conflict.

Resource conflict. Resource allocation problems are a major source of conflict in project management. Competition for resources, including personnel, tools, hardware, software, and so on, can lead to disruptive clashes among project members. The Triple-C approach can help secure resource cooperation.

Power conflict. Project politics lead to a power play which can adversely affect the progress of a project. Project authority and project power should be clearly delineated. Project authority is the control that a person has by virtue of his or her functional post. Project power relates to the clout and influence, which a person can exercise due to connections within the administrative structure. People with popular personalities can often wield a lot of project power in spite of low or non-existent project authority. The Triple-C model can facilitate a positive marriage of project authority and power to the benefit of project goals. This will help define clear leadership for a project.

Personality conflict. Personality conflict is a common problem in projects involving a large group of people. The larger the project, the larger the size of the management team needed to keep things running. Unfortunately, the larger management team creates an opportunity for personality conflicts. Communication and cooperation can help defuse personality conflicts. In summary, conflict resolution through Triple-C can be achieved by observing the following guidelines:

1. Confront the conflict and identify the underlying causes.
2. Be cooperative and receptive to negotiation as a mechanism for resolving conflicts.
3. Distinguish between proactive, inactive, and reactive behaviors in a conflict situation.
4. Use communication to defuse internal strife and competition.
5. Recognize that short-term compromise can lead to long term gains.
6. Use coordination to work toward a unified goal.
7. Use communication and cooperation to turn a competitor into a collaborator.

It is the little and often neglected aspects of a project that lead to project failures. Several factors may constrain the project implementation. All the relevant factors can be evaluated under the Triple-C model right from the project-initiation stage. The adaptation of the nursery rhyme below illustrates the importance of the little "things" in a project:

FALL OF THE KINGDOM

For the want of a nail, the horseshoe was lost;
For the loss of the horseshoe, the horse was lost;
For the loss of the horse, the rider was lost;
For the loss of the rider, the message was lost;
For the loss of the message, the battle was lost;
For the loss of the battle, the war was lost;
For the loss of the war, the Kingdom was lost.
All for the want of a nail!

FAILURE OF THE PROJECT

For the want of communication, the data was lost;
For the lack of the data, the information was lost;
For the lack of the information, the decision was lost;
For the lack of the decision, the cooperation was lost;
For the lack of the cooperation, the planning was lost;
For the lack of the planning, the coordination was lost;
For the lack of the coordination, the project failed.
All for the want of simple communication!

ALIGNMENT OF TRIPLE-C AND DEJI SYSTEMS MODEL WITH DMAIC

Many organizations now explore Six Sigma DMAIC methodology and associated tools to achieve better project performance. Six Sigma means six standard deviations from a statistical performance average. The Six Sigma approach allows for no more than 3.4 defects per million parts in manufactured goods or 3.4 mistakes per million activities in a service operation. To explain the effect of the Six Sigma approach, consider a process that is 99% perfect. That process will produce 10,000 defects per million parts. With Six Sigma, the process will need to be 99.99966% perfect in order to produce only 3.4 defects per million. Thus, Six Sigma is an approach that moves a process toward perfection. Six Sigma, in effect, reduces variability among products produced by the same process. By contrast, Lean approach is designed to reduce/eliminate waste in the production process.

Six Sigma provides a roadmap for the five major steps of DMAIC (Define, Measure, Analyze, Improve and Control), which are applicable to the planning and control steps of project management. Both DEJI Systems Model and the Triple-C model align well with DMAIC as summarized below:

Define, Measure, Analyze, Improve, and Control
Design, Evaluate, Justify, and Integrate
Communication, Cooperation, and Coordination

We cannot improve what we cannot measure. Triple-C provides a sustainable approach to obtaining cooperation and coordination for DMAIC during improvement efforts. DMAIC requires project documentation and reporting, which coincide with project control requirements.

This chapter has presented a general introduction to the Triple-C approach. The next three chapters are respectively devoted discussions of Communication, Cooperation, and Coordination. A summary of lessons to be inferred from a Triple-C approach are:

- Use proactive planning to initiate project functions.
- Use preemptive planning to avoid project pitfalls.
- Use meetings strategically. Meeting is not *work*. Meeting should be done to facilitate work.
- Use project assessment to properly frame the problem, adequately define the requirements, continually ask the right questions, cautiously analyze risks, and effectively scope the project.
- Be bold to terminate a project when termination is the right course of action. Every project needs an exit plan. In some cases, there is victory in capitulation.

The sustainability of the Triple-C approach is summarized below:

1. For effective communication, create good communication channels.
2. For enduring cooperation, establish partnership arrangements.
3. For steady coordination, use a workable organization structure.

BASIC FUNCTION OF PLANNING

Planning is a basic function of leadership. First, we discuss planning in the context of project planning, based on our earlier summation that every organizational pursuit can be couched as a project. Thus, project planning represents general planning within the context of organizational goals and objectives.

The important considerations in project feasibility studies are presented. Guidelines are presented for issuing or responding to requests for proposals. Budgeting is discussed as a planning tool for project efforts. Work breakdown structure (WBS) is discussed as it relates to getting more accurate work data for project plans so that appropriate communication can be initiated towards the end goal of securing cooperation, which then facilitates coordination. The overall theme of project planning should center on involving all organizational components and stakeholders. This is exactly what the Triple-C model was developed for.

PLANNING AND COMMUNICATION OBJECTIVES

The key to a successful project is good planning. Project planning provides the basis for the initiation, implementation, and termination of a project. It sets guidelines for specific project objectives, project structure, tasks, milestones, personnel, cost, equipment, performance, and problem resolutions. An analysis of what is needed and what is available should be conducted in the planning phase of new projects. The availability of technical expertise within the organization and outside the organization should be reviewed. Here are some guidelines for systems-wide project plans:

1. View a project plan as having tentacles that stretch across the organization and communication must be done across those tentacles.
2. Use project plans to coordinate across functional boundaries.
3. Establish plans as the platform over which project control will be done later on.
4. Leverage the diverse personalities and skills within the project environment.
5. Make room for contingent re-planning due to scope changes.
6. Empower workers to manage at the activity level.
7. Identify value-creating tasks and complementing activities.
8. Define specific milestones to facilitate project tracking.
9. Use checklists, tables, charts, and other visual tools project to communicate the plan.
10. Establish project performance metrics.

Although planning is a specific starting step in the project life cycle, it actually stretches over all the other steps of project management. Planning and re-planning permeate the project management life cycle. The major knowledge areas of project management, as presented by the Project Management Institute (PMI), are administered in a structured outline covering six basic clusters. The implementation clusters represent five process groups that are followed throughout the project life cycle. Each cluster itself consists of several functions and operational steps. When the clusters are overlaid on the knowledge areas, we obtain a two-dimensional matrix that spans the major process steps in the project. The monitoring and controlling clusters are usually administered as one lumped process group (monitoring and controlling). In some cases, it may be helpful to separate them to highlight the essential attributes of each cluster of functions over the project life cycle. In practice, the processes and clusters do overlap. Thus, there is no crisp demarcation of when and where one process ends and where another one begins over the project life cycle.

In general, project life cycle defines the following:

1. Resources that will be needed in each phase of the project life cycle.
2. Specific work to be accomplished in each phase of the project life cycle.

Project life cycle is distinguished from product life cycle. Project life cycle does not explicitly address operational issues whereas product life cycle is mostly about operational issues starting from the product's delivery to the end of its useful life. For technical projects, the shape of the life cycle curve may be expedited due to rapid developments that often occur in technology-based activities. For example, for a high technology project, the entire life cycle may be shortened, with a very rapid initial phase, even though the conceptualization stage may be very long.

Typical characteristics of project life cycle include the following:

1. Cost and staffing requirements are lowest at the beginning of the project and ramp up during the initial and development stages.
2. The probability of successfully completing the project is lowest at the beginning and highest at the end. This is because many unknowns (risks and uncertainties) exist at the beginning of the project. As the project nears its end, there are fewer opportunities for risks and uncertainties.
3. The risks to the project organization (project owner) are lowest at the beginning and highest at the end. This is because not much investment has gone into the project at the beginning, whereas much has been committed by the end of the project. There is a higher sunk cost manifested at the end of the project.
4. The ability of the stakeholders to influence the final project outcome (cost, quality, and schedule) is highest at the beginning and gets progressively lower toward the end of the project. This is intuitive because influence is best exerted at the beginning of an endeavor.
5. Value of scope changes decreases over time during the project life cycle while the cost of scope changes increases over time. The suggestion is to decide and finalize scope as early as possible. If there are to be scope changes, do them as early as possible.

The specific application context will determine the essential elements contained in the life cycle of the endeavor. Life cycles of business entities, products, and projects have their own nuances that must be understood and managed within the prevailing organizational strategic plan. The components of corporate, product, and project life cycles are summarized as follows:

Corporate (business) life cycle

Policy planning → Needs identification → Business conceptualization
→ Realization → Portfolio management

Product life cycle

Feasibility studies → Development → Operations → Product obsolescence

Project life cycle

Initiation → Planning → Execution → Monitoring and control → Closeout

In the initial stage of project planning, the internal and external factors that influence the project should be determined and given priority weights. Examples of internal influences on project plans include the following:

Infrastructure
Project scope
Labor relations
Project location
Project leadership
Organizational goal
Management approach
Technical personnel supply
Resource and capital availability

In addition to internal factors, a project plan can be influenced by external factors. An external factor may be the sole instigator of a project or it may manifest itself in combination with other external and internal factors. Such external factors include the following:

Public needs
Market needs
National goals
Industry stability
State of technology
Industrial competition
Government regulations

Project goals determine the nature of project planning. Project goals may be specified in terms of time (schedule), cost (resources), or performance (output). A project can be simple or complex. While simple projects may not require the whole array of project management tools, complex projects may not be successful without all the tools. Project management techniques are applicable to a wide collection of problems ranging from manufacturing to medical services.

The techniques of project management can help achieve goals relating to better product quality, improved resource utilization, better customer relations, higher productivity, and fulfillment of due dates. These can be expressed in terms of the following project constraints:

Performance specifications
Schedule requirements
Cost limitations

Project planning determines the nature of actions and responsibilities needed to achieve the project goal. It entails the development of alternate courses of action and the selection of the best action to achieve the objectives making up the goal. Planning determines what needs to be done, by whom, and when. Whether it is done for long-range (strategic) purposes or for short-range (operational) purposes, planning should be one of the first steps of project management.

SYSTEMS LEVELS OF PLANNING

Decisions involving strategic planning lay the foundation for the successful implementation of projects. Planning forms the basis for all actions. Strategic decisions may be divided into three strategy levels: supra-level planning, macro-level planning, and micro-level planning.

Supra-level planning: Planning at the supra-level deals with the big picture of how the project fits the overall and long-range organizational goals. Questions faced at this level concern potential contributions of the project to the welfare of the organization, its effect on the depletion of company resources, required interfaces with other projects within and outside the organization, risk exposure, management support for the project, concurrent projects, company culture, market share, shareholder expectations, and financial stability.

Macro-level planning: Planning decisions at the macro-level address the overall planning within the project boundary. The scope of the project and its operational interfaces should be addressed at this level. Questions faced at the macro-level include goal definition, project scope, availability of qualified personnel, resource availability, project policies, communication interfaces, budget requirements, goal interactions, deadline, and conflict resolution strategies.

Micro-level planning: The micro-level deals with detailed operational plans at the task levels of the project. Definite and explicit tactics for accomplishing specific project objectives are developed at the micro-level. The concept of management by objective (MBO) may be particularly effective at this level. MBO permits each project member to plan his or her own work at the micro-level. Factors to be considered at the micro-level of project decisions include scheduled time, training requirements, required tools, task procedures, reporting requirements, and quality requirements.

Project decisions at the three levels defined previously will involve numerous personnel within the organization with various types and levels of expertise. In addition to the conventional roles of the project manager, specialized roles may be developed within the project scope. Such roles include the following:

1. Technical specialist: This person will have responsibility for addressing specific technical requirements of the project. In a large project, there will typically be several technical specialists working together to solve project problems.
2. Operations integrator: This person will be responsible for making sure that all operational components of the project interface correctly to satisfy project goals. This person should have good technical awareness and excellent interpersonal skills.
3. Project specialist: This person has specific expertise related to the specific goals and requirements of the project. Even though a technical specialist may also serve as a project specialist, the two roles should be distinguished. A general electrical engineer may be a technical specialist on the electronic design components of a project. However, if the specific setting of the electronics project is in the medical field, then an electrical engineer with expertise in medical operations may be needed to serve as the project specialist.

PLANNING FOR COMMUNICATION

Planning is an ongoing process that is conducted throughout the project life cycle. Initial planning may relate to overall organizational efforts. This is where specific projects to be undertaken are determined. Subsequent planning may relate to specific objectives of the selected project. In general, a project plan should consist of the following components:

1. Summary of project plan: This is a brief description of what is planned. Project scope and objectives should be enumerated. The critical constraints on the project should be outlined. The types of resources required and available should be specified. The summary should include a statement of how the project complements organizational and national goals, budget size, and milestones.
2. Objectives: The objectives should be very detailed in outlining what the project is expected to achieve and how the expected achievements will contribute to the overall goals of a project. The performance measures for evaluating the achievement of the objectives should be specified.
3. Approach: The managerial and technical methodologies of implementing the project should be specified. The managerial approach may relate to project organization, communication network, approval hierarchy, responsibility, and accountability. The technical approach may relate to company experience on previous projects and currently available technology.
4. Policies and procedures: Development of a project policy involves specifying the general guidelines for carrying out tasks within the project. Project procedure involves specifying the detailed method for implementing a given policy relative to the tasks needed to achieve the project goal.

5. Contractual requirements: This portion of the project plan should outline reporting requirements, communication links, customer specifications, performance specifications, deadlines, review process, project deliverables, delivery schedules, internal and external contacts, data security, policies, and procedures. This section should be as detailed as practically possible. Any item that has the slightest potential of creating problems later should be documented.

6. Project schedule: The project schedule signifies the commitment of resources against time in pursuit of project objectives. A project schedule should specify when the project will be initiated and when it is expected to be completed. The major phases of the project should be identified. The schedule should include reliable time estimates for project tasks. The estimates may come from knowledgeable personnel, past records, or forecasting. Task milestones should be generated on the basis of objective analysis rather than arbitrary stipulations. The schedule in this planning stage constitutes the master project schedule. Detailed activity schedules should be generated under specific project functions.

7. Resource requirements: Project resources, budget, and costs are to be documented in this section of the project plan. Capital requirements should be specified by tasks. Resources may include personnel, equipment, and information. Special personnel skills, hiring, and training should be explained. Personnel requirements should be aligned with schedule requirements so as to ensure their availability when needed. Budget size and source should be presented. The basis for estimating budget requirements should be justified and the cost allocation and monitoring approach should be shown.

8. Performance measures: Measures of evaluating project progress should be developed. The measures may be based on standard practices or customized needs. The method of monitoring, collecting, and analyzing the measures should also be specified. Corrective actions for specific undesirable events should be outlined.

9. Contingency plans: Courses of actions to be taken in the case of undesirable events should be predetermined. Many projects have failed simply because no plans have been developed for emergency situations. In the excitement of getting a project under way, it is often easy to overlook the need for contingency plans.

10. Tracking, reporting, and auditing: These involve keeping track of the project plans, evaluating tasks, and scrutinizing the records of the project.

Planning for large projects may include a statement about the feasibility of subcontracting part of the project work. Subcontracting may be needed for various reasons including lower cost, higher efficiency, and logistical convenience.

MOTIVATING FOR COOPERATION

Motivation is an essential component of implementing project plans. National leaders, public employees, management staff, producers, and consumers may all need to be motivated about project plans that affect a wide spectrum of society. Those who will play active direct roles in the project must be motivated to ensure productive participation. Direct beneficiaries of the project must be motivated to make good use of the outputs of the project. Other groups must be motivated to play supporting roles to the project.

Motivation may take several forms. For projects that are of a short-term nature, motivation could be either impaired or enhanced by the strategy employed. Impairment may occur if a participant views the project as a mere disruption of regular activities or as a job without long-term benefits. Long-term projects have the advantage of giving participants enough time to readjust to the project efforts. Some of the essential considerations in aligning project plans for motivational purposes include the following elements:

Global coordination across functional lines
Balancing of task assignments
Goal-directed task analysis
Human cognitive information flow among the project team
Ergonomics and human factors considerations
Work load assessment considering fatigue, stress, emotions, sentiments, etc.
Interpersonal trust and collegiality
Project knowledge transfer lines
Harmony of personnel along project lines

Classical concepts of motivation suggest that management involves knowing exactly what workers are expected to do and ensuring that they have the tools and skills to do it well and cost effectively. This means that management requires motivating workers to get things done. Thus, successful management should be able to predict and leverage human behavior. An effective manager should be interested in both results and the people he or she works with. Whatever definition of management is embraced, it ultimately involves some human elements with behavioral and motivational implications. In order to get a worker to work effectively, he or she must be motivated. Some workers are inherently self-motivated, self-directed, and self-actuated. There are other workers for whom motivation is an external force that must be managerially instilled based on the two basic concepts of theory X and theory Y. Theory X assumes that the worker is essentially uninterested and unmotivated to perform his or her work. Motivation must be instilled into the worker by the adoption of external motivating agents. A theory X worker is inherently indolent and requires constant supervision and prodding to get him or her to perform. To motivate a theory X worker, a mixture of managerial actions may be needed. The actions must be used judiciously, based on the prevailing circumstances. Examples of motivation approaches under theory X are:

• Rewards to recognize improved effort
• Strict rules to constrain worker behavior
• Incentives to encourage better performance
• Threats to job security associated with performance failure

Theory Y assumes that the worker is naturally interested and motivated to perform his or her job. The worker views the job function positively and uses self-control and self-direction to pursue project goals. Under theory Y, management has the task of taking advantage of the worker's positive intuition so that his or her actions coincide with the objectives of the project. Thus, a theory Y manager attempts to use the worker's self-direction as the principal instrument for accomplishing work. In general, theory Y facilitates the following:

Worker-designed job methodology
Worker participation in decision making
Cordial management-worker relationship
Worker individualism within acceptable company limits

There are proponents of both theory X and theory Y and managers who operate under each or both can be found in any organization. The important thing to note is that whatever theory one subscribes to, the approach to worker motivation should be conducive to the achievement of the overall goal of the project.

LEVERAGING HIERARCHY OF NEEDS OF WORKERS

The needs of project participants must be taken into consideration in any project planning in accordance with the prevailing personal and behavioral landscape of the project. A common tool for accomplishing this is Maslow's hierarchy of needs, which was first introduced in Chapter 3 as part of the DEJI Model. With this framework, the following considerations are proposed:

1. Physiological needs: The needs for the basic things of life, such as food, water, housing, and clothing. *This is the level where access to money is most critical.*
2. Safety needs: The needs for security, stability, and freedom from threat of physical harm. *The fear of adverse environmental impact may inhibit project efforts.*
3. Social needs: The needs for social approval, friends, love, affection, and association. *For example, public service projects may bring about a better economic outlook that may enable individuals to be in a better position to meet their social needs.*
4. Esteem needs: The needs for accomplishment, respect, recognition, attention, and appreciation. *These needs are important not only at the individual level but also at the national level.*
5. Self-actualization needs: These are the needs for self-fulfillment and self-improvement. They also involve the availability of opportunity to grow professionally. *Work improvement projects may lead to self-actualization opportunities for individuals to assert themselves socially and economically. Job achievement and professional recognition are two of the most important factors that lead to employee satisfaction and better motivation.*

Hierarchical motivation implies that the particular motivation technique utilized for a given person should depend on where the person stands in the hierarchy of needs. For example, the need for esteem takes precedence over physiological needs when the latter are relatively well satisfied. Money, for example, cannot be expected to be a very successful motivational factor for an individual who is already on the fourth level of the hierarchy of needs. The hierarchy of needs emphasizes the fact that things that are highly craved in youth tend to assume less importance later in life.

There are two motivational factors classified as the *hygiene factors* and *motivators*. Hygiene factors are necessary but not sufficient conditions for a contented worker. The negative aspects of the factors may lead to a disgruntled worker, whereas their positive aspects do not necessarily enhance the satisfaction of the worker. Examples include the following:

1. *Administrative policies*: Bad policies can lead to the discontent of workers, while good policies are viewed as routine with no specific contribution to improving worker satisfaction.
2. *Supervision*: A bad supervisor can make a worker unhappy and less productive, while a good supervisor cannot necessarily improve worker performance.
3. *Worker conditions*: Bad working conditions can enrage workers, but good working conditions do not automatically generate improved productivity.
4. *Salary*: Low salaries can make a worker unhappy, disruptive, and uncooperative, but a raise will not necessarily provoke him to perform better. While a raise in salary will not necessarily increase professionalism, a reduction in salary will most certainly have an adverse effect on morale.
5. *Personal life*: Miserable personal life can adversely affect worker performance, but a happy life does not imply that he or she will be a better worker.
6. *Interpersonal relationships*: Good peer, superior, and subordinate relationships are important to keep a worker happy and productive, but extraordinarily good relations do not guarantee that he or she will be more productive.
7. *Social and professional status*: Low status can force a worker to perform at *his* or *her* level whereas high status does not imply performance at a higher level.
8. *Security*: A safe environment may not motivate a worker to perform better, but an unsafe condition will certainly impede productivity.

Motivators are motivating agents that should be inherent in the work itself. If necessary, work should be redesigned to include inherent motivating factors. Some guidelines for incorporating motivators into jobs are as follows:

1. *Achievement*: The job design should give consideration to opportunities for worker achievement and avenues to set personal goals to excel.
2. *Recognition*: The mechanism for recognizing superior performance should be incorporated into the job design. Opportunities for recognizing innovation should be built into the job.

3. *Work content*: The work content should be interesting enough to motivate and stimulate the creativity of the worker. The amount of work and the organization of the work should be designed to fit a worker's needs.
4. *Responsibility*: The worker should have some measure of responsibility for how his or her job is performed. Personal responsibility leads to accountability which invariably yields better work performance.
5. *Professional growth*: The work should offer an opportunity for advancement so that the worker can set his or her own achievement level for professional growth within a project plan.

The aforementioned examples may be described as job enrichment approaches with the basic philosophy that work can be made more interesting in order to induce an individual to perform better. Normally, work is regarded as an unpleasant necessity (a necessary evil). A proper design of work will encourage workers to become anxious to go to work to perform their jobs.

MANAGEMENT BY OBJECTIVE

To motivate people, we must know their personal characteristics so that we can apply appropriate leadership tools. MBO is the management concept whereby a worker is allowed to take responsibility for the design and performance of a task under controlled conditions. It gives workers a chance to set their own objectives in achieving project goals. Workers can monitor their own progress and take corrective actions when needed without management intervention. Workers under the concept of theory Y appear to be the best suited for the MBO concept. MBO has some disadvantages which include the possible abuse of the freedom to self-direct and possible disruption of overall project coordination. The advantages of MBO include the following:

1. It encourages workers to find better ways of performing their jobs.
2. It avoids over-supervision of professionals.
3. It helps workers become better aware of what is expected of them.
4. It permits timely feedback on worker performance.

MANAGEMENT BY EXCEPTION

Management by exception (MBE) is an after-the-fact management approach to control. Contingency plans are not made and there is no rigid monitoring. Deviations from expectations are viewed as exceptions to the normal course of events. When intolerable deviations from plans occur, they are investigated, and then an action is taken. The major advantage of MBE is that it lessens the management workload and reduces the cost of management. However, it is a dangerous concept to follow especially for high-risk technology-based projects. Many of the problems that can develop in complex projects are such that after-the-fact corrections are expensive or even impossible. As a result, MBE should be carefully evaluated before adopting it. The previously described motivational concepts can be implemented successfully for specific large projects. They may be used as single approaches or in a combined strategy. The motivation approaches may be directed at individuals or groups of individuals, locally or at the national level.

FEASIBILITY DRIVES COOPERATION

If a project does not pass the test of feasibility, it cannot be successful in the cooperation and coordination efforts. The feasibility of a project can be ascertained in terms of technical factors, economic factors, or both. A feasibility study is documented with a report showing all the ramifications of the project.

Technical feasibility: Technical feasibility refers to the ability of the process to take advantage of the current state of the technology in pursuing further improvement. The technical capability of the personnel as well as the capability of the available technology should be considered.

Managerial feasibility: Managerial feasibility involves the capability of the infrastructure of a process to achieve and sustain process improvement. Management support, employee involvement, and commitment are key elements required to ascertain managerial feasibility.

Economic feasibility: This involves the feasibility of the proposed project to generate economic benefits. A benefit-cost analysis and a break-even analysis are important aspects of evaluating the economic feasibility of new industrial projects. The tangible and intangible aspects of a project should be translated into economic terms to facilitate a consistent basis for evaluation.

Financial feasibility: Financial feasibility should be distinguished from economic feasibility. Financial feasibility involves the capability of the project organization to raise the appropriate funds needed to implement the proposed project. Project financing can be a major obstacle in large multiparty projects because of the level of capital required. Loan availability, credit worthiness, equity, and loan schedule are important aspects of financial feasibility analysis.

Cultural feasibility: Cultural feasibility deals with the compatibility of the proposed project with the cultural setup of the project environment. In labor-intensive projects, planned functions must be integrated with local cultural practices and beliefs. For example, religious beliefs may influence what an individual is willing to do or not to do.

Social feasibility: Social feasibility addresses the influences that a proposed project may have on the social system in the project environment. The ambient social structure may be such that certain categories of workers may be in short supply or non-existent. The effect of the project on the social status of the project participants must be assessed to ensure compatibility. It should be recognized that workers in certain industries may have certain status symbols within the society.

Safety feasibility: Safety feasibility is another important aspect that should be considered in project planning. Safety feasibility refers to an analysis of whether the project is capable of being implemented and operated safely with minimal adverse effects on the environment. Unfortunately, environmental impact assessment is often not adequately addressed in complex projects.

Political feasibility: A politically feasible project may be referred to as a "politically correct project." Political considerations often dictate the direction for a proposed project. This is particularly true for large projects with national visibility that may have significant government inputs and political implications. For example, political necessity may be a source of support for a project regardless of the project's merits. On the other hand, worthy projects may face insurmountable opposition simply because of political factors. Political feasibility analysis requires an evaluation of the compatibility of project goals with the prevailing goals of the political system.

BUDGET PLANNING FOR COOPERATION

After the planning for a project has been completed, the next step is the allocation of the resources required to implement the project plan. This is referred to as budgeting or capital rationing. Budgeting is the process of allocating scarce resources to the various endeavors of an organization. It involves the selection of a preferred subset of a set of acceptable projects due to overall budget constraints. Budget constraints may result from restrictions on capital expenditures, shortage of skilled personnel, shortage of materials, or mutually exclusive projects. The budgeting approach can be used to express the overall organizational policy. The budget serves many useful purposes including the following:

- Performance measure
- Incentive for efficiency
- Project selection criterion
- Expression of organizational policy
- Plan of resource expenditure
- Catalyst for productivity improvement
- Control basis for managers and administrators
- Standardization of operations within a given horizon

The preliminary effort in the preparation of a budget is the collection and proper organization of relevant data. The preparation of a budget for a project is more difficult than the preparation of budgets for regular and permanent organizational endeavors. Recurring endeavors usually generate historical data which serve as inputs to subsequent estimating functions. Projects, on the other hand, are often onetime undertakings without the benefits of prior data. The input data for the budgeting process may include inflationary trends, cost of capital, standard cost guides, past records, and forecast projections. Budget data collection may be accomplished by one of several available approaches including top-down budgeting and bottom-up budgeting.

COOPERATION AND PROJECT WORK BREAKDOWN STRUCTURE

WBS represents a family tree hierarchy of project operations required to accomplish project objectives. It is particularly useful for purposes of planning, scheduling, and control. Tasks that are contained in the WBS collectively describe the overall project. The tasks may involve physical products (e.g., steam generators), services (e.g., testing), and data (e.g., reports, sales data). The WBS serves to describe the link between the end objective and the operations required to reach that objective. It shows work elements in the conceptual framework for planning and controlling. The objective of developing a WBS is to study the elemental components of a project in detail. It permits the implementation of the "divide and conquer" concepts. Overall project planning and control can be improved by using a WBS approach. A large project may be broken down into smaller subprojects which may, in turn, be systematically broken down into task groups.

Individual components in a WBS are referred to as WBS elements and the hierarchy of each is designated by a level identifier. Elements at the same level of subdivision are said to be of the same WBS level. Descending levels provide increasingly detailed definition of project tasks. The complexity of a project and the degree of control desired determine the number of levels in the WBS. Each WBS component is successively broken down into smaller details at lower levels. The process may continue until specific project activities are reached. The basic approach for preparing a WBS is as follows:

Level 1: It contains only the final project purpose. This item should be identifiable directly as an organizational budget item.
Level 2: It contains the major subsections of the project. These subsections are usually identified by their contiguous location or by their related purposes.
Level 3: It contains definable components of the level 2 subsections.

Subsequent levels are constructed in more specific detail depending on the level of control desired. If a complete WBS becomes too crowded, separate WBSs may be drawn for the level 2 components. A WBS summary should normally accompany the WBS. This is a narrative of the work to be done. It should include the objectives of the work, its nature, resource requirements, and tentative schedule. Each WBS element is assigned a code that is used for its identification throughout the project life cycle. Alphanumeric codes may be used to indicate element level as well as component group.

COMMUNICATION AND INFORMATION FLOW

Information flow is very crucial in project planning. Information is the driving force for project decisions. The value of information is measured in terms of the quality of the decisions that can be generated from the information. What appears to be valuable information to one user may be useless to another. Similarly, the timing of information can significantly affect its decision-making value. The same information that is useful in one instance may be useless in another. Some of the crucial factors affecting the value of information include accuracy, timeliness, relevance, reliability, validity, completeness, clearness, and comprehensibility. Proper information flow in project management ensures that tasks are accomplished when, where, and how they are needed. Information starts with raw data (e.g., numbers, facts, specifications). The data may pertain to raw material, technical skills, or other factors relevant to the project goal. The data is processed to generate information in the desired form. The information feedback model acts as a management control process that monitors project status and generates appropriate control actions. The impact of the information on the project goal is routed back through an information feedback loop. The feedback information is used to improve the next cycle(s) of the operation. The final information output provides the basis for improved management decisions. The key questions to ask when requesting, generating, or evaluating information for project management are as follows:

What data are needed to generate the information?
Where are the data going to come from?
When will the data be available?
Is the data source reliable?
Are there enough data?
Who needs the information?
When is the information needed?
In what form is the information needed?
Is the information relevant to project goals?
Is the information accurate, clear, and timely?

As an example, the information flow model described before may be implemented to facilitate the inflow and outflow of information linking several functional areas of an organization, such as the design department, manufacturing department, marketing department, and customer relations department. The lack of communication among functional departments has been blamed for many of the organizational problems in industry. The use of a standard information flow model can help alleviate many communication problems. The information flow model can be expanded to take into account the uncertainties that may occur in the project environment.

COMPLEXITY OF COMMUNICATION

Communication complexity increases with an increase in the number of communication channels. It is one thing to wish to communicate freely, but it is another thing to contend with the increased complexity when more people are involved in the communication. The statistical formula of combination can be used to estimate the complexity of communication as a function of the number of communication channels or number of participants. The combination formula is used to calculate the number of possible combinations of r objects from a set of n objects. This is written as

$$_nC_r = \frac{n!}{r![n-r]!}$$

In the case of communication, for illustration purposes, we assume communication is between two members of a team at a time. That is, combination of two from n team members. That is, number of possible combinations of two members out of a team of n people. Thus, the formula for communication complexity reduces to the expression as follows, after some of the computation factors cancel out:

$$_nC_2 = \frac{n(n-1)}{2}$$

In a similar vein, Badiru (2008) introduced a formula for cooperation complexity based on the statistical concept of permutation. Permutation is the number of possible arrangements of k objects taken from a set of n objects. The permutation formula is written as

$$_nP_k = \frac{n!}{(n-k)!}$$

Thus, for the number of possible permutations of two members out of a team of n members is estimated as

$$_nP_2 = n(n-1)$$

Permutation formula is used for cooperation because cooperation is bi-directional. Full cooperation requires that if A cooperates with B, then B must cooperate with A. But A cooperating with B does not necessarily imply B cooperating with A.

TRIPLE-C AND CONFLICT RESOLUTION

When implemented as an integrated process, the Triple-C model can help avoid conflicts in a project. When conflicts do develop, it can help in resolving the conflicts. Several sources of conflicts can exist in large projects. Some of these are discussed next.

Schedule conflict: Conflicts can develop because of improper timing or sequencing of project tasks. This is particularly common in large multiple projects. Procrastination can lead to having too much to do at once, thereby creating a clash of project functions and discord among project team members. Inaccurate estimates of time requirements may lead to infeasible activity schedules. Project coordination can help avoid schedule conflicts.

Cost conflict: Project cost may not be generally acceptable to the clients of a project. This will lead to project conflict. Even if the initial cost of the project is acceptable, a lack of cost control during project implementation can lead to conflicts. Poor budget allocation approaches and the lack of a financial feasibility study will cause cost conflicts later on in a project. Communication and coordination can help prevent most of the adverse effects of cost conflicts.

Performance conflict: If clear performance requirements are not established, performance conflicts will develop. Lack of clearly defined performance standards can lead each person to evaluate his or her own performance based on personal value judgments. In order to uniformly evaluate quality of work and monitor project progress, performance standards should be established by using the Triple-C approach.

Management conflict: There must be a two-way alliance between management and the project team. The views of management should be understood by the team. The views of the team should be appreciated by management. If this does not happen, management conflicts will develop. A lack of a two-way interaction can lead to strikes and industrial actions which can be detrimental to project objectives. The

Triple-C approach can help create a conducive dialogue environment between management and the project team.

Technical conflict: If the technical basis of a project is not sound, technical conflicts will develop. New industrial projects are particularly prone to technical conflicts because of their significant dependence on technology. Lack of a comprehensive technical feasibility study will lead to technical conflicts. Performance requirements and systems specifications can be integrated through the Triple-C approach to avoid technical conflicts.

Priority conflict: Priority conflicts can develop if project objectives are not defined properly and applied uniformly across a project. Lack of a direct project definition can lead each project member to define his or her own goals which may be in conflict with the intended goal of a project. Lack of consistency of the project mission is another potential source of priority conflicts. Over-assignment of responsibilities with no guidelines for relative significance levels can also lead to priority conflicts. Communication can help defuse priority conflicts.

Resource conflict: Resource allocation problems are a major source of conflict in project management. Competition for resources, including personnel, tools, hardware, software, and so on, can lead to disruptive clashes among project members. The Triple-C approach can help secure resource cooperation.

Power conflict: Project politics lead to a power play which can adversely affect the progress of a project. Project authority and project power should be clearly delineated. Project authority is the control that a person has by virtue of his or her functional post. Project power relates to the clout and influence which a person can exercise due to connections within the administrative structure. People with popular personalities can often wield a lot of project power in spite of low or non-existent project authority. The Triple-C model can facilitate a positive marriage of project authority and power to the benefit of project goals. This will help define clear leadership for a project.

Personality conflict: Personality conflict is a common problem in projects involving a large group of people. The larger a project, the larger the size of the management team needed to keep things running. Unfortunately, the larger management team creates an opportunity for personality conflicts. Communication and cooperation can help defuse personality conflicts.

In summary, conflict resolution through Triple-C can be achieved by observing the following guidelines:

1. Confront the conflict and identify the underlying causes.
2. Be cooperative and receptive to negotiation as a mechanism for resolving conflicts.
3. Distinguish between proactive, inactive, and reactive behaviors in a conflict situation.
4. Use communication to defuse internal strife and competition.
5. Recognize that short-term compromise can lead to long-term gains.
6. Use coordination to work toward a unified goal.
7. Use communication and cooperation to turn a competitor into a collaborator.

INTRODUCTION TO ENTERPRISE BUSINESS ACUMEN (EBA)

Enterprise-wide strategies are of great importance for the survival of any organization. In this book, we introduce the concept of Enterprise Business Acumen (EBA) to formalize the desirable characteristics of a leader with respect to organizational insight, judgment, wisdom, and institutional knowledge. Enterprise, in this context, covers the various functions and challenges in an organization dedicated to developing and executing operational imperatives. Information technology, budgeting, human resources, marketing, operating culture, innovation management, and infrastructure are some of the types of operational imperatives of an organization. Business, in the context of EBA, refers to the wide scope of the priorities of the organization that impinge upon profitability and survival. The concept of EBA and can be integrated into the leadership evaluative matrix of AMALA, introduced earlier. Leaders can be evaluated and rated on the basis of their performance with EBA metrics.

INTRODUCTION TO AGGREGATED MATRIX ASSESSMENT OF LEADERSHIP ATTRIBUTES (AMALA)

Leaders have diverse roles in any organization. But, how do we assess and rate leaders on a comparative basis, devoid of subjective personality influences? This is answered by an adaptive assessment model, introduced in this book, as aggregated matrix assessment of leadership attributes (AMALA). Mandatory, desirable, optional, and/or elective attributes that can be included in the matrix assessment are technical skills, communication effectiveness, team-oriented abilities, professional outlook, organizational skills, equity and diversity skills, empathy, and sympathy qualities. The AMALA matrix lists the desired attributes as a comparative basis in the first column of the matrix, against the names (or representations) of competing leaders. A possible generic layout of the matrix is presented below.

Leadership Attributes	Leader1 (Name)	Leader2 (Name)	Leader3 (Name)	Leader4 (Name)	Leader5 (Name)
Technical skills	√	√			√
Communication effectiveness		√		√	
Team-oriented	√		√		
Professional	√	√			√
Organized		√		√	
DEI skills	√				
Empathy			√		
Sympathy	√	√			√

A user can include the attributes of interest, pertinent for the prevailing needs of the organization. A weighted averaging scheme can be developed to aggregate the attributes for cases where the attributes don't carry equal weights. In regard to the application of EBA and AMALA, the following are some of the desirable attributes of good leaders:

- Ability to negotiate with subordinates and superiors
- Ability to lead with positivity
- Ability to break problems down into manageable chunks
- Ability to improvise
- Ability to communicate and articulate strategies
- Ability to make decisions based on facts rather than based on emotions
- Ability to exude optimism
- Ability to adapt
- Ability to demonstrate perseverance
- Ability to absorb criticism and more forward
- Ability to persuade vertically and horizontally
- Ability to delegate
- Ability to seek help when needed
- Ability to collaborate irrespective of differences
- Ability to sacrifice ego for righteousness

All of these attributes, as well as other to be determined by the organization, can form the comparative elements in the AMALA matrix. Customization of EBA and AMALA to the specific needs of organizations is a required for their effectiveness.

THE ABILENE PARADOX AND THE TRIPLE-C

A classic example of conflict in project planning is illustrated by the *Abilene Paradox* (Harvey, 1974). The text of the paradox, as presented by Harvey, is summarized below.

It was a July afternoon in Coleman, a tiny Texas town. It was a hot afternoon. The wind was blowing fine-grained West Texas topsoil through the house. Despite the harsh weather, the afternoon was still tolerable and potentially enjoyable. There was a fan blowing on the back porch, there was cold lemonade, and finally there was entertainment: dominoes. Perfect for the conditions. The game required little more physical exertion than an occasional mumbled comment, "shuffle 'em," and an unhurried movement of the arm to place the spots in the appropriate position on the table. All in all, it had the makeup of an agreeable Sunday afternoon in Coleman until Jerry's father-in-law suddenly said, "Let's get in the car and go to Abilene and have dinner at the cafeteria."

Jerry thought, "What, go to Abilene? Fifty-three miles? In this dust storm and heat? And in an un-airconditioned 1958 Buick?" But Jerry's wife chimed in with, "Sounds like a great idea. I'd like to go. How about you, Jerry?" Since Jerry's own preferences were obviously out of step with the rest, he replied, "Sounds good to me," and added, "I just hope your mother wants to go."

"Of course I want to go," said Jerry's mother-in-law. "I haven't been to Abilene in a long time." So into the car and off to Abilene they went. Jerry's predictions were fulfilled. The heat was brutal. The group was coated with a fine layer of dust that was cemented with perspiration by the time they arrived. The food at the cafeteria provided first-rate testimonial material for antacid commercials.

Some four hours and 106 miles later, they returned to Coleman, hot and exhausted. They sat in front of the fan for a long time in silence. Then, both to be sociable and to break the silence, Jerry said, "It was a great trip, wasn't it?" No one spoke. Finally, his father-in-law said, with some irritation, "Well, to tell the truth, I really didn't enjoy it much and would rather have stayed here. I just went along because the three of you were so enthusiastic about going. I wouldn't have gone if you all hadn't pressured me into it."

Jerry couldn't believe what he just heard. "What do you mean 'you all'?" he said. "Don't put me in the 'you all' group. I was delighted to be doing what we were doing. I didn't want to go. I only went to satisfy the rest of you. You're the culprits." Jerry's wife looked shocked. "Don't call me a culprit. You and Daddy and Mama were the ones who wanted to go. I just went along to be sociable and to keep you happy. I would have had to be crazy to want to go out in heat like that."

Her father entered the conversation abruptly. "Hell!" he said. He proceeded to expand on what was already absolutely clear. "Listen, I never wanted to go to Abilene. I just thought you might be bored. You visit so seldom I wanted to be sure you enjoyed it. I would have preferred to play another game of dominoes and eat leftovers in the icebox."

After the outburst of recrimination, they all sat back in silence. There they were, four reasonably sensible people who, of their own volition, had just taken a 106-mile trip across a godforsaken desert in a furnace-like temperature through a cloud-like dust storm to eat unpalatable food at a hole-in-the-wall cafeteria in Abilene, when none of them had really wanted to go. In fact, to be more accurate, they'd done just the opposite of what they wanted to do. The whole situation simply didn't make sense. It was a paradox of agreement.

This example illustrates a problem that can be found in many organizations or project environments. Organizations often take actions that totally contradict their stated goals and objectives. They do the opposite of what they really want to do. For most organizations, the adverse effects of such diversion, measured in terms of human distress and economic loss, can be immense. A family group that experiences the Abilene paradox would soon get over the distress, but for an organization engaged in a competitive market, the distress may last a very long time. Six specific symptoms of the paradox are identified by Harvey (1974).

1. Organization members agree privately, as individuals, as to the nature of the situation or problem facing the organization.
2. Organization members agree privately, as individuals, as to the steps that would be required to cope with the situation or solve the problem they face.
3. Organization members fail to accurately communicate their desires and/or beliefs to one another. In fact, they do just the opposite and, thereby, lead one another into misinterpreting the intentions of others. They misperceive the collective reality. Members often communicate inaccurate data (e.g., "Yes, I agree"; "I see no problem with that"; "I support it") to other members of the organization. No one wants to be the lone dissenting voice in the group.

4. With such invalid and inaccurate information, organization members make collective decisions that lead them to take actions contrary to what they want to do and, thereby, arrive at results that are counterproductive to the organization's intent and purposes. For example, the Abilene group went to Abilene when it preferred to do something else.

5. As a result of taking actions that are counterproductive, organization members experience frustration, anger, irritation, and dissatisfaction with tier organization. They form subgroups with supposedly trusted individuals and blame other subgroups for the organization's problems.

6. The cycle of the Abilene paradox repeats itself with increasing intensity if the organization members do not learn to take responsibility for their agreement.

We have witnessed many project situations where, in private conversations, individuals express their discontent about a plan and yet fail to repeat their displeasure in a group setting. Consequently, other members are never aware of the dissenting opinions … and the collective plan may fail. In large organizations, the Triple-C model can help in managing communication, cooperation, and coordination functions to avoid the Abilene paradox, or, at least, mitigate it. The lessons to be learned from proper approaches to project planning, communication, cooperation, and coordination can help avoid unwilling trips to Abilene.

REFERENCES

Badiru, Adedeji B. (1987) "Communication, Cooperation, Coordination: The Triple C of Project Management," in *Proceedings of 1987 IIE Spring Conference*, Washington, DC, May 1987, pp. 401–404.

Badiru, Adedeji B. (2008) Triple C Model of Project Management. Taylor & Francis, CRC Press, Boca Raton, FL.

Badiru, Adedeji B. (2019) Project Management: Systems, Principles, and Applications, Second Edition. Taylor & Francis/CRC Press, Boca Raton, FL.

Badiru, Adedeji B., Foote, Bobbie L., Leemis, Larry, Ravindran, Anurachalam, and Williams, Larry (1993) Recovering from a Crisis at Tinker Air Force Base. PM Network, 7(2), 10–23.

Harvey, Jerry B. (1974) The Abilene Paradox: The Management of Agreement. Organizational Dynamics, 3, 63–80. https://doi.org/10.1016/0090-2616(74)90005-9

5 Leadership Case Study
HFM of Nigeria

THE CASE OF HONEYWELL FLOUR MILLS

Having presented the basic background of organizational leadership with an industrial engineering framework, we now present the case of study of how Mr. Tunde Odunayo organized and led Honeywell Flour Mills (HFM), Lagos, Nigeria, from the ground up, until his retirement in 2014 (Badiru, 2015; Badiru et al, 2001). His departure from the organization sent the company into a leadership tailspin that eventually led to the demise of the company in 2021. His success in creating the company from scratch is credited by the authors to his natural practice of many of the principles of industrial engineering, even though he was not an industrial engineer. The authors further proclaim that the company's demise could have been averted if his successors had harnessed the power behind the industrial engineering principles that Tunde enacted during his leadership tenure.

Tunde Odunayo joined Honeywell in 1992 in the position of Group Managing Director and moved to Victoria Island. There were eight companies in all as listed below:

1. Honeywell Enterprises Limited
2. Honeywell Construction Company Limited
3. Poly Ventures Limited
4. Coral Products Limited
5. Skyview Estates Limited
6. Pivot Engineering Company Limited
7. Honeywell Fisheries Limited
8. Pavilion Technology Limited

After four years in the position, the Group profitability objectives were far from being realized, and a panacea was not readily in sight. How was the Group to proceed? In early 1997, the Chairman, Mr. Oba Otudeko set up a committee of three Senior Executives comprised of the Group Managing Director, Director of Human Resources, and General Manager, Business Planning to examine all aspects of the Group operations, including structure, strategy, manning. Their task was to come up with a blueprint for taking the Group business to greater heights. The report was submitted in January 1997.

The Chairman found it difficult to accept the report. He promptly called a meeting of all senior executives to his Idi-Ishin residence in Ibadan where free from interruptions, they could all discuss the report. The Group Consultant, Mr. B. T. Oluokun, was invited to the meeting. What was the crux of the report?

DOI: 10.1201/9781003311348-5

The Chairman was cumbered by the report's findings indicating the Group was not in the right businesses, some of which should be consequently exited. It also reported the Group was not structured in a manner that would assist it to fulfill its business objectives. Each company was recommended to have its own board, and competent CEOs should be appointed for each company so that Group management could be decentralized. If those steps were taken, the positions of Group Managing Director, Group Director of Human Resources, and Group Financial Controller were not required and should be taken out. How could this type of report be submitted by a committee of three persons, two of whom would be losing their positions if the report was accepted? This was the Chairman's difficulty with the report.

Tunde Odunayo, Group Managing Director and Chairman of the Committee, was, on the other hand, very happy with the report. He slept most soundly in the night following the submission of the report. After an exhaustive consideration of the committee's report, the Chairman accepted it for adoption.

The Chairman then called a smaller meeting of Messrs. Oluokun, Tunde Odunayo, and himself to another living room in his residence. He expressed his appreciation to Tunde for the courage of the report, and offered him a new position. He offered him the position of Managing Director of Honeywell Flour Mills Limited, a new company which would commence flour milling operations after building construction and equipment installation. The company would be sited at the Tin Can Island Port. Tunde thanked the Chairman for the new offer and asked for some days to think about it.

In the next several years that followed, many of the businesses listed in the preceding paragraphs were discontinued, and today, of those eight companies, only Skyview Estates Limited (now rebranded as Uraga Estates Limited), Pivot Engineering Company Limited, and Pavilion Technology Limited remain in operations.

New businesses were started over the course of the next five years. They are:

- Honeywell Flour Mills Limited
- Honeywell Oil & Gas Limited
- Honeywell Superfine Foods Limited

Honeywell Superfine Foods Limited was later merged with Honeywell Flour Mills Plc.

This company had been registered in 1985 as Gateway Honeywell Flour Mills Limited. A name-change to Honeywell Flour Mills Limited was secured in 1995 when the decision was made to start the flour milling business. Mr. Tunde Odunayo assumed the position of pioneer Managing Director in April 1997, and moved his office to Tin Can Island Apapa, there to complete the building construction works which had started in 1995. Equipment installation was completed in July 1998 and the first product sale took place on July 13, 1998.

The success story of the company, under the leadership of Tunde Odunayo, will require to be comprehensively researched by a Business School in order to, one day, use its strategies to develop new educational case studies. Suffice it to mention that installed capacity grew from 200 metric tonnes per day to 2,610 metric tonnes per day between 1998 and 2013, while Tunde was at the helm. The company grew from a very small flour mill to a large-sized mill and reached the pinnacle of respect by its peers. Its brands not only became very well known by consumers all over Nigeria, but also became the brands of first choice.

WHAT DID THE COMPANY DO TO THRIVE?

- Honeywell Superfine Flour
- Honeywell Semolina
- Honeywell Spaghetti
- Honeywell Pasta
- Honeywell Noodles
- Honeywell Wheat Meal

By successfully producing these six wheat products, Honeywell Flour Mills became the competition, the corporate brand to defeat within its industry.

The company instituted a rigorous wheat quality specification and procurement process, plus a painstaking production quality control process, ensuring that the product offerings were the best in the market place. Honeywell became the preferred brand, and ended up controling 70% of the market share for wheat meal. For other products, the company took the second position in terms of market share, not because it had conceded the first position to the competition, but because the company still required to ramp up its production capacity so that it could offer more volumes of the other brands to the teeming population.

THE CREATION OF BRANDS

Soon after the company commenced business in 1998, and event that was clearly marked by the dispatch of the first truck load of product to the customer (that product was bran), it became clear that the installed capacity of 200 metric tonnes per day was inadequate to meet the growing demand for a quality of flour that had never been experienced thus far by the Nigerian Baking Industry. The milling capacity was consequently increased by the installation of a new Mill B in year 2000, after about 12 months of project planning, equipment manufacturing by Buhler AG Switzerland. This investment increased installed milling capacity to 610 metric tonnes per day.

The new milling capacity was soon taken up fully by customers, and a full investment payback (ROI) was achieved within 15 months. An increasingly unfulfilled level of demand for the Honeywell Superfine Flour led the way to yet another milling capacity increase, and by June 2005 a milling capacity of 1000 metric tonnes per day had been added, giving a total milling capacity of 1610 metric tonnes per day, a 700% increase in milling capacity within a relatively short period of 7 years.

A significant aspect of the capacity expansion projects was that they were conceived, designed, and installed under the project leadership of the management, a group of professionals who were not formally trained in industrial engineering. The projects encompassed factory building reconstruction, sometimes reconstruction from the grounds up. In the subsequent expansion projects, entire new factory buildings had to be designed and constructed.

Performance budgetary targets which were set by the Board did not make allowances for project management responsibilities of the management. Budgetary targets and the projects completion timetable were strictly monitored by the Board, and it would not be an exaggeration if the inference was made here that the management was always under pressure to meet these expectations.

It was in the hot crucible of high performance expectations, supported by a stick and carrot approach of the company ownership, that a fine and seasoned management was baked. And with this, a high performing organization underlined by great consumer brands was also raised. Thus, between July 1998 and March 2014, over a period of 16 financial reporting years, it was only in two reporting periods that the performance targets were not met.

The balance sheet of the company grew from N536 million in April 1998 to N55 billion in March 2013. The March 2014 financial statements went through external validation by the auditors and, subsequently, received approval of the Nigerian Stock Exchange before the results were released to the general public.

Of course, the budgeting approval process created its own challenges. As performance expectations were met from year to year, targets were set even higher. A high level of capacity utilization did not create significant room for volume growth, so profit targets had to be significantly met from improving gross margins and ratios, until additional capacities were added. Gross margins were significantly determined by the mix of international wheat prices and local flour prices. These two factors were not fully in control of management, so year-on-year profit improvements were difficult to achieve. Placing them into the budget was the biggest challenge faced by the management.

More often than not, it was never possible to fully recover cost increases from adjustments to flour prices. Regardless of these realities, budget approval proceedings were always tension-filled. A reward-based performance management system was strongly raising performance expectations, and management struggled to have expectations placed at manageable levels.

On one hand, the unspoken concern of the owners was not to set targets that management would find easy to reach and thereby claim productivity bonus for targets that perhaps did not stretch the management. On the other hand, the management relied on analyzing the fundamentals of the business to show what possibilities were achievable, given the business fundamentals, the economic environment, and the activities of the competitors in the market place.

On balance, the profit targets negotiation sessions were beneficial to both parties, and some of those benefits can be enumerated as follows:

- The productivity and performance level of management was greatly enhanced.
- Shareholder value was on the ascent on a yearly basis.
- Operational processes were under constant improvement, as internal efficiency was a key objective.
- The turnaround period of 24 hours, being the maximum allowable customer waiting period between payment for goods and dispatch of the order, endeared the company to its customers. Amongst other strategies, this was beneficial to customer repeat orders and sales volumes.

- The constantly improving internal processes soon became a good training platform for staff. Within a decade, a management succession pipeline had been developed.
- Staff benefitted from numerous local and overseas training courses, sometimes for extended periods of three to six months. Overseas training of more than 12 months is known to have been achieved. Training brought about tremendous improvements to staff and corporate performance.

GREAT BRANDS WERE CREATED

First of all, the mission statement had clearly established the business objective and the type of corporate brand that should be developed:

To produce high quality flour brands for our highly esteemed customers through the effort of our highly motivated workforce.

The quality of the corporate brand emerged from this mission statement. The staff also modeled the brand and were encouraged to internalize the brand values, namely quality, integrity, and leadership, among others. The company's products characterized and communicated the brand values, and with support from well-crafted advertising communication, the most admired brands in its categories were created.

Honeywell Superfine Flour was and remains a successful business-to-business (B-to-B) brand but the launching of the following consumer products helped to position the Honeywell brands as the consumers' brands of first choice across the country:

Honeywell Semolina	2006
"O Noodles"	2008 (Re-launched as Honeywell Noodles in 2010)
Honeywell Spaghetti	2009
Honeywell Pasta	2009
Honeywell Wheat Meal	2009

STRATEGIC ADVERTISING

Some marketing schools of thought would argue that "advertising creates brands." Rather, Tunde believed that advertising only supports the aspiration of great brands, but does not, by itself, create the brands. Great brands aspire to be noticed, to be in the hands of every consumer or buyer within its field of target. And great advertising can support these aspirations.

A poorly conceived product does not become a great brand because of the supporting advertising campaign. A good product, well designed to meet the needs and aspirations of a carefully identified segment of the market, and standing taller than competing products in the market in terms of quality, utility, appeal, and affordability, is more likely to succeed in the market than a brand that comes in with poorly researched market entry parameters and poorly engineered characteristics.

Great brands are created by entrepreneurs, owners, managers, or some other creative persons who conceptualize the products and design them, by brand owners who define the target consumer market, fit them within the relevant market segments, and prepare the strategies for their commercialization. Microsoft, Coca-Cola, and Apple's iPad became great brands because of these things and more. Advertising was used to support the aspirations of these brands. Where brand aspiration meets with consumer aspiration, then success is at the door. Of course, all these steps require appropriate financial resources to bring great brand plans into reality.

Suffice it to say here that Honeywell Flour Mills went to great lengths to develop strong advertisements and advertisement campaigns to support its great brands. And all of the company's brands are great brands!

The iPod Man advert, also known as "bam bam la la," a name curled from its chorus, is one of the finest commercial campaigns ever launched on Nigerian television. It is loved by all, young and old, despite the fact that Honeywell Noodles the brand for which the campaign was created was targeted at persons in year 3–22 age bracket. For a tasty Noodle product like Honeywell Noodles, a tremendous rise in sales volume was the outcome.

Under Odunayo's leadership, one of the several complimentary proclamations goes as follows: "Today Honeywell Flour Mills brands are well used throughout the country and they are the toast of the consumers. The flour brand commands 20% of the market, the Semolina brand has 12% (limited only by production volume), the Wheat Meal has 70% of the market. The spaghetti and pasta products command 10% of the market. Honeywell Noodles takes second place in the market limited only by its production capacity, but production expansion plans will soon widen its reach in the market, until it can one day aspire to beat the giant in its category."

Deliberately, names of competing brands are not mentioned here as they shall not be glorified or put down in this case study. A well-researched marketing analysis of brands in these categories will be a good place to read about other competing brands and their place in relation to the Honeywell brand leaders of that time.

ENTRY INTO THE Nigerian STOCK EXCHANGE

Honeywell Flour Mills began from a humble beginning. It was a business fully owned by Mr. Obafoluke Otudeko, a former banker at Cooperative Bank, Ibadan who followed his early passion to establish himself in business after a flourishing career in the bank. A Fellow of the Institute of Bankers and of the Association of Certified and Corporate Accountants, you can say that he was well prepared to proceed to a business career.

This resoundingly successful entrepreneur who prefers to be known simply as Oba Otudeko is an enigma. Apart from his visibility in the pursuit of his business objectives, very little is publicly known about his intellect, his energy, his resourcefulness, his punishing daily schedule, looks that belie his age, his power of recall of names, faces, events and scores and scores of financial numbers. Very little can also be said of him here. Nor of his numerous business interests and wealth. Suffice it to mention that he was the architect of the Honeywell Group of businesses, the one who was already seeking to recruit a Group Managing Director for his eight companies at the age of 49 years.

In the course of Oba Otudeko's Presidency of the Nigerian Stock Exchange, an internal decision was made to take one of his companies to the Nigerian Stock Exchange. Honeywell Flour Mills was the obvious target. The preparations began in earnest in 2008 and by 2009, the exercise had been successfully wrapped up after receiving all regulatory approvals.

The internal arrangements for this purpose included the 100% acquisition of Honeywell Superfine Foods Limited (manufacturers of Noodles and Pasta) by Honeywell Flour Mills, thus making it a full subsidiary of the company. At the conclusions of the capital market outing, the paid up capital of the company increased from N1billion to N3.965billion with a share premium of N6.462billion.

Through the unlocking of value that the platform of the capital market provided, the shareholders' funds increased from N5.19billion in 2008 to N14.275billion at the end of the 2009 calendar year. The fully paid shares also increased from 1billion shares of N1 each, to 7,930,197,658 shares of 50 kobo each.

The combined Offer for Sale/Subscription was for 941,176,472 ordinary shares of 50 kobo each at N8.50 per share. The application list closed on December 31, 2008, and the transactions were completed and approved later in 2009.

By March 2013, through yet again an internal restructuring, Honeywell Superfine Foods Limited was merged by absorption into Honeywell Flour Mills Plc after obtaining all regulatory and court approvals.

Ever the consummate spokesperson for Honeywell Flour Mills, Mr. Odunayo never missed an opportunity to pitch the company's products anywhere, anytime, anyhow. He did several live radio and television appearances, and gave many newspaper interviews as well. During the Honeywell 2008 IPO (Initial Public Offer), he went on a nationwide television campaign to sell stocks of the company and encourage potential shareholders to buy Honeywell shares. To further promote the company, Tunde Odunayo appeared on "Sunrise," a morning program of channels TV, Lagos.

ABICS CONSULTING SERVICES FOR HONEYWELL FLOUR MILLS

The prime author's long-term professional relationship with Honeywell Flour Mills goes back several years. The sustainable relationship has thrived and strengthened over the years. One culmination of the relationship is evidenced by the good news of ISO 9001:2000 certification of Honeywell as conveyed in the email exchange that is echoed below.

From: Tunde Odunayo
Sent: Thursday, January 26, 2006 8:38 AM
To: abadiru@abicsnet.com
Subject: ISO 9001:2000 CERTIFICATION

Dear Prof,

I am pleased to inform you that Honeywell Flour Mills will be officially presented with the ISO 9001:2000 certificate tomorrow at the Muson Centre by the Standards Organisation of Nigeria -the representative of the International Standards Organisation in Nigeria.

Of course, the journey started after the ABICS seminar in 2002, at the end of which we set up a Quality Steering Committee under Dr. Ogunmoyela to take us towards this route. It was hard work for the committee, especially upon the company's insistence that it should be a wholly in-house effort without the use of external consultants and based on the fact that we have adequate human resources to do it by ourselves. The impetus from the ABICS seminar is very well appreciated.

The presentation will receive adequate press coverage over the next 4 weeks, as it is significant that after 50years of flour milling in Nigeria. Honeywell Flour Mills is the first Flour Mill in Nigeria to receive an ISO certification for its Quality Management Systems.

In year 2005, we had also received an NIS award (NIGERIAN INDUSTRIAL STANDARD) for the quality of our brand. The ISO 9001:2000 award is far more significant, as it is not only an international award, it is also a mark of distinction for our management systems covering production, new product development, marketing, HR, administration and management information reporting systems.

Best Regards,

Tunde

ABICS (A. B. International Consulting Services) is an industrial training and consultancy business administered by the prime author, Adedeji Badiru and his wife, Iswat, occasionally run out of their home as a professional outreach to Nigerian organizations. Other clients of ABICS have included oil and gas companies in Nigeria, Russia, Canada, and Mexico. Badiru's email reply to the good news is presented below:

From: Adedeji Badiru
Sent: Thursday, January 26, 2006 9:23 AM
To: Tunde Odunayo
Subject: RE: ISO 9001:2000 CERTIFICATION

Dear MD:

CONGRATULATIONS on this extraordinary achievement. This is an important leadership lesson for all the corporate entities in Nigeria, indeed, in the whole of Africa. This would not have been possible without your selfless and persistent leadership of Honeywell Flour Mills. You are a leader with a unique vision for your organization.

Thanks for giving the incipient credit to ABICS. Our own mode of client service is to get organizations started so that they can proceed internally with their project management and quality improvement goals. Endeavours like this have a greater potential for success if internal ownership is cultivated so that employees can collectively carry out the mission effectively. The journey may be painful initially (as you observed), but the eventual rewards justify the agony of starting. Honeywell Flour Mills has just confirmed this axiom of corporate operation. I applaud you for leading your company way ahead of its industry peers. I wish you more success. Tony and I are still interested in writing a corporate case study on you and your company. When time and schedule permit, we should embark on that aim.

ABICS and our associates celebrate with you (remotely) on this exceptional achievement of ISO 9001:2000 Certification. We only wish we had been invited to attend the event in our physical presence. If there is anything else that our group can do for Honeywell Flour Mills, please let me know. We are presently offering training in Six Sigma techniques.

Please convey our regards and congratulations to all the employees, colleagues, and friends at Honeywell Flour Mills.

Cheers!
Deji Badiru
Jan. 26, 2006

The two American consultants who accompanied Adedeji on the trip to Nigeria for the Honeywell training seminar mentioned above were Mr. Tony Mayfield and Mr. Jason Chandler. It was, indeed, Mr. Mayfield, impressed with what he observed at Honeywell, who first floated the idea of writing a corporate case study on Mr. Odunayo and his exemplary leadership of Honeywell Flour Mills in 2002. It took several years before the festschrift on Tunde Odunayo was published in 2015 (Badiru, 2015). The reprint of portions of festschrift in this book has taken several additional years to accomplish (2002 to 2015 to 2023).

ABICS subsequently expanded into guide book publications, which facilitated the compilation of the 2015 festschrift. In addition to offering customized training programs, ABICS now publishes reference books for home, work, and leisure.

LEADERSHIP BY EXAMPLE

Tunde Odunayo encouraged every staff member to imbibe the core values of the company. Integrity, Can-do Spirit, leadership at all levels, and respect for all were the foremost core values around which the fabric of the organization was woven. It can be said of the Honeywell Flour Mill staff that Integrity and Can do Spirit constituted the basic character of all staff, just as the brands deliver on their promise of quality and integrity. Monthly management seminars reinforced these values and helped to cement them into the culture of the organization.

Under Mr. Odunayo's leadership, Honeywell Flour Mills, Plc won several awards, accolades, and international recognition for its products and managerial accomplishments.

LEADERSHIP TRANSITION

Three months to clocking the age of 60 years, on October 4, 2012, Tunde gave notice to the Chairman of his intention to retire after 22 years of service, 15 years of which had been served at that time as the pioneer CEO of Honeywell Flour Mills Plc.

Amongst other things, he said the following in his Notice of Retirement:

I thank God for His grace which has taken the company to the performance heights that it has attained in the last 15 years. I have also been most privileged to work in the Honeywell Group under your leadership. Your strong engagement across the socio-economic landscape of Nigeria, and in particular your entrepreneurial spirit have become for me the driving example, as I executed the mandate that you gave me in the years that have passed.

"... You provided me a job which became for me a significantly enriching life's experience.

The job has grown tremendously, 15 years have elapsed as the pioneer CEO of the company; I am 60 years old and a younger hand is now required to steer the ship. It is a beautiful ship with a good team, its lights shine radiantly into the horizon and I am very proud...."

"The time has come to look in the horizon for what may quietly await me as I move on to this new phase of my life. I am grateful to God for bringing me thus far in all the colors of personal and professional experiencing."

"Long live Honeywell Group, Long live Honeywell Flour Mills Plc"

The Chairman, Oba Otudeko in accepting Tunde's Notice of Retirement, thanked him for his long service, his success at placing the company's brands in the awareness of Nigerians throughout the country, for making them the consumers' first choice of brands in the product categories that they belong to. Above all, he thanked him for producing consistently superior corporate performance throughout the years in service.

Tunde was persuaded by the Chairman and the board to remain in the position of Executive Vice Chairman/CEO for another 15 months until a rigorous valida-tion of the successor pipeline was completed by an international consulting outfit. As a recognition of his meritorious service, and his contributions to the growth and success of the company, Tunde had earlier been elevated to the position of Executive Vice Chairman in year 2009 at the time of the Public Offer of the com-pany's shares.

Tunde articulated reasons for the move which took many by surprise, which had nothing to do with his age, but everything to do with the people who worked for him, a bona fide demonstration of good leadership. Only his management team had been aware of the impending move, and even they were not fully versed in Tunde's dedica-tion to cultivating his followers for growth and increased responsibility. Having men-tored so many of his junior colleagues from trainee positions to senior management positions, he reasoned it was high time he stepped aside to give them opportunities for self-actualization. Some of them who were recruited soon after their graduation, and having served upward of 18 to 20 years, were already grown far beyond their middle age. The danger of overstaying is that by the time a CEO finally decides to leave, or is compelled by circumstances and or by nature to take a bow, some of the smartest well-trained subordinates may already have taken their exit in order to look for other self-actualization opportunities. Tunde was eager to avoid this situation.

He further reasoned that a successful CEO is the one that leaves at the right time after implementing a good succession arrangement; he is the one who leaves no management capability gap at the time of his exit. For those reasons, and given the company was in a continuous path of flight and ascent, he concluded the time was ripe. The younger ones would bring internal knowledge, experience, energy, and fresh motivation to extend the frontier of success in managing the business.

On March 31, 2014 after 17 years as pioneer CEO, he bowed out gracefully from the company at the end of a Valedictory Board Meeting which took place on that day in his honor.

The public announcement by the Board of his retirement stated, among other things, alia that ".... during his tenure as Chief Executive Officer, Tunde Odunayo exhibited exemplary leadership, commitment and drive in building an enterprise that is very well known for its superior quality brands and sustained corporate performance. His investment in developing leaders and a high performing team has been confirmed by a rigorous selection process which led to the appointment of three senior management staff to Executive positions, to the Board of the Company....."

The new Managing Director was formerly the Commercial Director at Honeywell Superfine Foods, a subsidiary which became fully merged with the company in March 2013. He had been seconded to that company from Honeywell Flour Mills in 2009 from where he received tutelage and grooming. He had joined Honeywell Flour Mills 20 years earlier as a young accountant. He celebrated his 50th birthday a few months before his new appointment.

Other appointments that were made from the internal succession pipeline were as follows:

- **Divisional Managing Director.** He was a former University lecturer and Head of the Department of Soil Science. He had bagged a first-class degree in the same discipline before proceeding to the University of Cranfield in the UK for his PhD degree. He joined the company 16 years ago from a university teaching position and was given an intensive business training, as well overseas technical training in wheat milling technology. He had earlier risen to the position of Production Director before this new appointment. He will assume responsibility for the noodles and pasta business based in Ikeja.
- **Executive Director, Supply Chain.** (He joined the company 20 years ago as a Management Trainee, and was later to pioneer the leadership of the logistics and supply chain function. He has gone past his mid-40s.
- **Executive Director Marketing.** He joined the company eight years ago as General Manager, Marketing after successful marketing careers at Procter & Gamble and Cadbury. He has gone past his mid-40s.

A LEADER OF DIVERSE INTERESTS

Still young at heart and looks, as a demonstration that he is not all work and no play, Mr. Odunayo, upon retirement, was able to carve out some free time to travel to Brazil for 2014 FIFA Soccer World Cup. About his Brazil travel journal, it is interesting to note that some go to Brazil to sing and dance, some go to Brazil to take in the sights, some go to Brazil to eat and drink, some go to Brazil to dream and fantasize, and some go to Brazil to seek out new opportunities. Mr. Odunayo truly went to Brazil to unwind from the arduous corporate world and to be a real soccer fan!

In its heydays, HFM developed nationally acclaimed promotional items (see Figure 5.1) that made it readily identifiable and respected in the Nigeria food industry.

FIGURE 5.1   Industry leadership legacy of Honeywell Flour Mills (before 2021).

EXAMPLE FROM POLITICAL LEADERSHIP

Having completed the details of the HFM case study, it is worth recognizing one last aspect of Tunde's leadership that strengthened his positive results. Over the 22 years that he served Honeywell Flour Mills, Tunde gave thousands of talks and passed along as many communications in the form of memos and emails. Tunde utilized tenets from Triple-C and DEJI Systems Model to formulate his messages. Similarly and not coincidentally, the newly elected governor of the Osun State of Nigeria delivered a public address echoing similar tenets of these two successful IE models. Details of that address follow.

What follows is the inauguration address delivered by the governor of Osun State of Nigeria, Ademola Adeleke, on November 27, 2022:

1. Today, I stand on the podium of history with a sacred pledge to God and the good people of Osun State that my intellect, passion, and strength will be devoted to nothing but the welfare, peace, and security of our people.
2. As I accept the mantle of leadership entrusted upon me by the people of our dear State, I am conscious of the enormous challenges and responsibilities ahead of me and I will, beginning from this hour, work day and night with a deep sense of purpose to be a servant to you all.
3. Our gathering here to witness the materialization of a new era is not an accident of history. We are all programmed by the Almighty God to be what we have been, what we are and what we will be in future.
4. Let me use this occasion to salute the founding fathers of our dear state, all the past administrators and Governors of Osun state as well as everyone who has contributed one way or the other to the growth, stability and progress of our state in the last 31 years.
5. I equally salute our fallen heroes both at the national and state level for the great sacrifice they have made to have a country and a state to call our own.
6. I am well aware of the fact that my responsibility as the Governor and Chief Security Officer of Osun State entails meeting the legitimate expectations of our people. Therefore, I promise that those expectations of the workers, traders, artisans, farmers, business owners, students, pensioners, traditional

and religious leaders and indeed all residents of Osun State will be met by the grace of God and the cooperation of everyone.

7. Under my watch as the Governor of Osun State, I will boldly correct all past injustice, corrupt acts or policies by any previous administration which are against the collective interest of our people.

8. Let me state here that from the education, health, mining Sector, agriculture, road infrastructure and supply of portable water, let it be known to all that it is no longer going to be business as usual. And I repeat, it is no longer business as usual.

9. Our administration will demonstrate a high sense of urgency, transparency, justice and innovation to tackle and solve the problems of poverty, illiteracy, disease, and poor infrastructure.

10. Your Governor will be a people's Governor. I will be accessible, responsive, consultative and proactive in handling small and big matters of State Governance.

11. I know that as a product of the collective will of you my people, there is a heavy weight of history on my shoulders and I accept the urgency of your expectations, the depth of your aspirations and your conviction in me to build a better State.

Development agenda for Osun State

12. Ladies and Gentlemen, you will recall that our party, the Peoples Democratic Party campaigned on a five point agenda namely:
    a. Welfare of workers and pensioners;
    b. Boosting the state's economy
    c. Home-grown infrastructure policy;
    d. People-focused policy on education, affordable Health Care, Security and Social Welfare
    e. Agro-Based Industrialization for Wealth and Job Creation.

13. While a detailed program of action will soon be unveiled on each of these 5-point agenda, let me quickly give you some insights into our policy direction.

Education

14. It is disheartening to see our State at the bottom of the national educational ratings especially in public primary and Secondary schools examinations. My administration will launch reform with direct focus on improvement of learning environment and outcome. Our target is to reverse the poor performance of students in public examinations within the next few years.

15. To achieve this target, we will prioritize in-service training and welfare of teachers, enhancement of school environment, entrenchment of discipline in the school system as well as involvement of the Parents-Teachers Association in our school administration system.

16. Our administration will soon convene an emergency Education Conference to articulate our blueprint on the restoration of the state's education glory.

Agro-industrial

17. We regard as critical the need to boost wealth creation, job opportunities and food security through innovative agricultural reforms covering all the agricultural value chains.

18. Aside the introduction of modern agricultural practices, we will shift attention to agriculture for export to take advantage of global markets.
19. While we target agriculture export earnings, we will establish modern farmers' markets where producers and off takers can transact businesses.

Tourism

20. In our desire to diversify Osun economy, my administration will target the tourism sector to boost our GDP and create new jobs.
21. Osun State is the historical capital of the Yoruba people and it is my intention to develop a strong business model to transform this rich historical heritage into huge tourism. We will encourage and partner with the traditional institutions, business organizations and foreign partners to develop the culture and tourism industry which includes our beautiful waterfalls, the Osun Osogbo World Heritage site and many cultural festivals across the state.

Women and youth development

22. Women and youths are important segments in our development program. I had earlier set up a youth advisory council which has produced an impressive blueprint for youth development in Osun State. I plan to ensure speedy implementation of their recommendations. More importantly, let me assure our women that our administration will be gender – sensitive in all appointments.

Health

23. The health sector in Osun State is in need of urgent attention. Our government will give premium attention to improvement of the Primary Health care services. We will improve the working conditions of health workers and expand the coverage of our health insurance scheme. We plan to establish a standard diagnostic center using the Public Private Partnership model and adopt measures that will retain medical practitioners in Osun state.

Climate change and ICT

24. Our administration will collaborate with Development partners to address problems of environmental pollution and Climate Change. We will equally promote digital literacy, tech innovations and create opportunities for our teeming youths in the ICT sector.

Local government reforms

25. In line with our campaign manifesto, our government will ensure local government autonomy in line with the provisions of the Nigeria constitution.
26. We will restore the lost glory of our local governments in order to make them more responsive to the needs of the people at the grassroots.
27. All policies that are not favorable to the growth of effective local government administration will be reviewed in accordance with the law.
28. No form of illegality will be allowed to stand and all acts of impunity committed by the immediate past administration on local government administration will be reversed following due process.

The public service

29. My administration will restore the integrity of the civil service which has been bastardized through favoritism and political considerations.
30. We will give the civil service a better orientation with a view to restoring its professionalism. In doing this, we will be fair and firm as we act only in pursuance of public interest.
31. Let me therefore announce an immediate return to status quo of all fresh appointments, placements and other major decisions taken by the immediate past administration with effect from the 17th of July, 2022.
32. I wish to assure all labor unions in the state of our administration's willingness to protect workers interest and promote their welfare at all times. We will be a labor friendly government.

Security

33. We will embark on a security sector reform that will target crime prevention, detection, neighborhood policing and better synergy among security agencies. The Amotekun Corps will be strengthened while our administration will ensure operational linkage between local hunters and the Amotekun Corps. A "Know Your Neighbor" security initiative will be implemented. By the grace of God, very soon, Osun will return to its old nature of being a haven of peace.

Executive orders

34. My Good People of Osun, since you elected me as your Governor on the 16th of July 2022, which the INEC announced on July 17th 2022, the former Governor, Alh Isiaka Gboyega Oyetola maliciously started putting road blocks to make things difficult and almost impossible for the new administration to serve you.
35. Mass employment were carried out without budgetary provisions for salary payments for the new employees; even when the state was struggling to pay salaries and deliver other services. Various hurried and criminally backdated Contracts were awarded and again without budgetary provisions.
36. Appointments of several Obas were hurriedly done without following due process, just to mention a few.
37. All efforts to get the Governor set up a Transition Committee and submit hand-over notes in line with best practices proved abortive.
38. It is therefore my desire to ask for your patience and understanding to give my administration a few weeks to review and sort out all the actions and malicious confusions which the immediate past administration has created since July 17th 2022. Those actions were indeed vindictive measures against the people of Osun State for voting them out of office.
39. Consequently, I hereby issue the following directives which will be backed up with appropriate Executive orders:

    Immediate freezing of all government accounts in banks and other financial institutions. Immediate establishment of a panel to carry out an inventory and recover all government assets. An immediate establishment of a panel to review all appointments and major decisions of the immediate past administration taken after the 17th July, 2022

An immediate reversal to the constitutionally recognized name of our state, Osun State. All government insignia, correspondences and signages should henceforth reflect Osun state rather than State of Osun which is unknown to the Nigerian constitution.

Appreciation

40. As I conclude this address, let me express my appreciation to the President and Commander-In-Chief of the Armed Forces of the Federal Republic of Nigeria, His Excellency Muhammadu Buhari, GCFR for allowing a conducive atmosphere for a peaceful, free and fair election in Osun state.

41. I thank the national leadership of our party, the Peoples Democratic Party under the able leadership of our Chairman, His Excellency Sen. Iyorchia Ayu and his National Working Committee for their immeasurable support.

42. I also thank our presidential candidate, His Excellency Alhaji Atiku Abubakar, GCON, Vice presidential candidate His Excellency Sen. Ifeanyi Okowa, CON, all PDP state Governors, the Minority caucuses of the Senate, House of Representatives and Osun State House of Assembly and other party stakeholders who stood by Osun state to reclaim our mandate which was stolen four years ago.

43. Let me also thank the state chapter of the PDP under the able leadership of Dr. Adekunle Akindele, the immediate past leadership under Hon. Sunday Bisi and all party leaders in the 30 local government areas and Ife East area office. I thank you for your steadfastness, loyalty and commitment to the ideals which the Peoples Democratic Party represents.

44. My deep sense of appreciation goes to my immediate family members beginning with my elder brother Dr, Adedeji Adeleke, my immediate elder sister Yeye Modupe Adeleke-Sanni, my wives Titi and Ngozi, my children, nephews and nieces especially David Adedeji Adeleke A.K.A. Davido for their love, support and sacrifice towards ensuring that we got this far. May your sacrifices never be in vain. I assure you that the selfless services of our late father Sen. Raji Ayoola Adeleke, our mother Prophetess Esther Adeleke and our late scion of the Adeleke family, the first Governor of Osun state, His Excellency/Alhaji Isiaka Adetunji Adeleke will always guide my actions as I serve the good people of Osun state.

45. I must not end this address without appreciating the Nigerian judiciary and the Independent National Electoral Commission (INEC) for their uprightness, neutrality, professionalism and courage in protecting our democracy without fear or favor.

46. I commend our security agencies for providing a peaceful and conducive atmosphere before, during and after the electoral process.

47. Finally, I extend my thanks to party men and women, friends, associates, supporters across party lines, the Media and the courageous Osun voters who have stood resolutely to give and defend this mandate. I am most sincerely grateful.

Conclusion

48. Ladies and Gentlemen, I have a grand vision of a new Osun state. But I cannot do it alone. I therefore seek the support and cooperation of everyone to make this happen.

49. I hereby extend a strong hand of fellowship to the other arms of governments, the Legislature and the Judiciary. As a former lawmaker and a graduate of Criminal Justice, I appreciate the importance of collaboration among the Executive, the Legislature and the Judiciary.

50. I call on all political parties to unite in the best interest of the state. Election is over. Now is the time for governance. We are open to fresh ideas in line with our manifesto.

51. For the purpose of emphasis, I will be a Governor for all Osun people regardless of differences in language, faith, political affiliation or any other considerations.

52. Welcome to a new Osun State, The State of The Living Spring.

53. Thank you and God bless you all.

As a closing exercising for readers, it is of interest to evaluate the effectiveness of this political talk with respect to the tools and techniques that have been presented throughout this book, including the following:

Triple-C Model
DEJI Systems Model
Enterprise Business Acumen (EBA)
Aggregated Matrix Assessment of Leadership Attributes (AMALA)

In the end, the passage of time and political tenure will determine if the governor delivers on the various promises and pledges contained in the well-crafted public talk. It is a fact that the realities of talks and actions help to distinguish great leaders, such as it did for Mr. Tunde Odunayo.

As a matter of how leadership matters, we conclude this case study by providing a selected summary of Tunde Odunayo's Leadership style:

- Personnel-oriented
- Customer-focused
- Taking ownership of decisions
- Being accountable for results (good or bad)
- Continual learning
- Inquisitiveness about work process
- Readiness to take action
- Frugal econometrics
- Firmness on commitments
- Continual business growth
- Readiness to engage
- Know when to leave

REFERENCES

Badiru, Adedeji (2015) Folaranmi Babatunde Odunayo – Festschrift in Honour of a Business Leader: Silent Works, Great Works. ABICS Publications, Dayton, OH.

Badiru, A. B., Omitaomu, H. Olufemi and Odunayo, F. B. (2001) "Design of a Matrix Knowledge Base Framework for Plant Maintenance Planning and Control," *Proceedings of the Maintenance and Reliability Conference 2001*, Gatlinburg, Tennessee, May 19–21, 2001.

Index

Printed in the United States
by Baker & Taylor Publisher Services